What Industry Leaders are Saying . . .

"This is a vital book for any company trying to get its entire organization behind its rapid product development efforts. It presents an extremely practical how-to approach. The insights on managing the front end of the process are excellent. Clearly, the most useful book I've read on this subject."

Christopher C. Cole, Group Vice President
Cincinnati Milacron

"Pro/ENGINEER lets us link mechanical design with downstream functions such as tooling, manufacturing, and packaging. This allows multifunctional teams to drive product development. By sharing a common 3D solid model, all team members can work in parallel. Suggested changes can be incorporated quickly using the system's associative and parametric capabilities. As a result, we've reduced our review cycle from weeks to days."

Steve Gompertz, Business Systems Project Leader
Medtronic, Inc.

"This book provides important insights on further accelerating the product development process. Future success depends on exceeding customer expectations, and this requires responding quickly to customers with new, exciting, high-quality products."

Lewis B. Campbell, Vice President
General Motors Corporation

"With Pro/ENGINEER, we are creating digital models more quickly and accurately than ever before. These models are crucial to meeting our product quality, cost, and time-to-market goals because they form the core of our entire development cycle."

Dick Allen, Vice President of Research and Development
Rubbermaid

"We have used the financial models in Chapter 2 to guide tradeoffs for our new TDS400A, which resulted in satisfied customers, a financially successful product and project, and rapid development. The tools in this book make the new Tektronix time-to-market goals a reality."

John Hengeveld, New Product Development Program Manager
Tektronix

Pro/ENGINEER® from Parametric Technology Corporation

With its commercial introduction in 1988, Pro/ENGINEER revolutionized the CAD/CAM industry by making solids modeling a viable technique for mechanical design. Since then, paradigm shifts such as concurrent engineering and the "virtual" organization, coupled with an expanding global marketplace, have increased pressure on manufacturers to reduce cycle times, lower costs and maximize quality.

To meet these challenges, PTC provides more than 70 fully associative software products designed to enhance a company's competitiveness by improving its product development processes. This includes capabilities in industrial design; mechanical design, including large-assembly management; functional simulation; manufacturing; and information management. Together, these Pro/ENGINEER Solutions provide the most comprehensive, tightly integrated product development environment possible.

This book is provided with the compliments of . . .

DEVELOPING PRODUCTS IN HALF THE TIME

PRESTON G. SMITH
DONALD G. REINERTSEN

VAN NOSTRAND REINHOLD
I(T)P™ A Division of International Thomson Publishing Inc.

New York • Albany • Bonn • Boston • Detroit • London • Madrid • Melbourne
Mexico City • Paris • San Francisco • Singapore • Tokyo • Toronto

Published by Van Nostrand Reinhold, a division of
International Thomson Publishing Inc.
The ITP logo is a trademark under license

The management techniques described in this book must be appropriately tailored by users to fit their specific needs.

Printed in the United States of America

For more information, contact:

Van Nostrand Reinhold
115 Fifth Avenue
New York, NY 10003

Chapman & Hall GmbH
Pappelallee 3
69469 Weinheim
Germany

Chapman & Hall
2-6 Boundary Row
London
SE1 8HN
United Kingdom

International Thomson Publishing Asia
221 Henderson Road #05-10
Henderson Building
Singapore 0315

Thomas Nelson Australia
102 Dodds Street
South Melbourne, 3205
Victoria, Australia

International Thomson Publishing Japan
Hirakawacho Kyowa Building, 3F
2-2-1 Hirakawacho
Chiyoda-ku, 102 Tokyo
Japan

Nelson Canada
1120 Birchmount Road
Scarborough, Ontario
Canada M1K 5G4

International Thomson Editores
Campos Eliseos 385, Piso 7
Col. Polanco
11560 Mexico D.F. Mexico

This is a customized, complimentary, non-returnable edition of Developing Products in Half the Time, ISBN 0-442-025092.

If you wish to purchase copies of the standard edition please contact Customer Service at 1-800-842-3636. Reference ISBN 0-442-02064-3.

Contents

10

Monitoring and Controlling Progress 169

11

Capacity Planning and Resource Allocation 189

12

Managing Risk 207

13

The Product–Process Duo 223

Foreword

I have the unique opportunity to be writing this foreword with twenty-twenty hindsight. My colleagues and I at Black & Decker repeatedly refer to *Developing Products in Half the Time* as we accelerate our development process. The practical tools from this guide have all passed the test of actual application. The results we have achieved demonstrate why thousands of companies in Europe, Asia, and North America are using this book to accelerate their product development.

We are proud of the success we have achieved at Black & Decker. On average, our cycle time has decreased by 50 percent on a broad assortment of power tool projects. Consequently, we think we have learned a lot about developing quality products quickly. It is hard to find anything we are doing to improve cycle time that does not appear in this book.

We have seen the dramatic payoff of approaches presented in this guide including:

- An accountable, dedicated, cross-functional team with a strong leader and management sponsorship
- Doing the up-front "homework" well and getting sharp early product definition
- A strong market orientation
- A sense of urgency with relentless schedule critical path management

These keys to our success are all clearly explained in this book, but the book doesn't stop with these techniques. Here you will find tools that range from the surprisingly intuitive, such as co-location, to more sophisticated ones, including product architecture and capacity management.

We have received excellent payback for the time invested in learning and implementing these techniques of cycle compression. Although we are pleased with our progress to date, we are not standing still—because we know that our competitors are not standing still either. Nor have we exhausted what this book can teach us. These pages cover other innovative tools that we have yet to apply consistently to their full potential. They include the techniques of incremental innovation and product architecture, as well as the management of capacity and risk.

We followed the advice in *Developing Products in Half the Time* by introducing incremental improvements in our product development process as well as starting several pilot projects. This approach is much faster than a large reengineering project and minimizes risk. At the same time, we trained the teams in the soft as well as hard skills required in the new environment we envisioned. Once we neared completion on the pilot projects, we undertook a reengineering process to clone our learning and best practices. We used the techniques outlined in Chapter 9, Achieving Overlapping Activities, to change our product development process from a sequential hand-off Tollgate System to a continuous flow concurrent engineering system that we call Milestones.

In addition to extending the use of these tools, we are rolling this program out to an ever-broadening circle at Black & Decker. We took the lead in our power tools product development centers. I will provide one indication of where we are headed: for the past few years we have been assigning a full-time manufacturing engineer to each major product development team. This has helped immensely in designing from the outset for manufacturability and in bridging the gap with our worldwide manufacturing plants. However, we have learned that this is not good enough. Now we have transferred design engineers to the plant, and people in the plant are studying this book so that they can apply the tools of rapid product development earlier, thus eliminating more cycle time.

When I speak at product development conferences, I suggest this book as a valuable guide. If you are one of those to whom it has yet to be suggested, you will be struck by some of the powerful concepts here that receive little attention elsewhere. For example, the economic analysis tools in Chapter 2 are without equal in simplicity and applicability. In addition, this book was describing how to tackle major opportunities in the Fuzzy Front End before most developers even knew what the term meant.

One of the things I like about this book—in contrast with others on the topic—is that although it is a book about speed, it does not recommend speed at any price. The authors begin by emphasizing that time to market has its price and that we should approach speed as we do any purchase: if the benefits outweigh the price, we buy it; if not, we don't.

This rational approach toward reducing cycle time has influenced many areas in Black & Decker. For example, because delay has a specific price, we look at how we run our support services much differently than we did when we considered them "efficient" cost centers with adequate queues to keep the workload level.

We are proud of the progress we have been able to make on time to market by using these tools. Yet, we also know that more work lies ahead. As our business becomes more global, new brand lines are added, and new technologies appear in our products, our development process will naturally become more complex. Unless we continue to apply and refine the kinds of tools you are about to learn in this book, our development process will consequently slow down. Because these tools have helped us achieve our current proficiency, and because Smith and Reinertsen are continually adding to and honing the tools, I trust that this new edition of their book will help us overcome the hurdles of the future too.

As you face time to market challenges, I highly recommend this book to you. I suggest reading it at least once; your competition probably has already!

Mike Brennan
Vice President, Product Development
Black & Decker

Preface

It has been almost five years since the hardcover version of this book was published. Since then we have continued to learn about what works to shorten development cycles and what doesn't work. Much of this knowledge comes from practitioners who use the hardcover version of the book and now number in the tens of thousands.

With this base of experience we have made some changes in this paperback edition. To make it easier to find key points in the chapters we have added the following icons:

Tool for accelerating development or measuring progress

Warning about pitfalls

A real-world example

We have also added Chapter 16 which presents some key observations based on our experiences over the last five years working with companies to shorten development cycles. At the end of this final chapter we have included a Reminder List which summarizes the most common performance improvement opportunities.

The content of the book has proved quite durable. We have encountered no areas where we feel the original message was erroneous. This is not due to unusual foresight on our part, but rather to the way the book was created—we started by observing what was working and documenting it, not by creating a theory and looking for justification. You will find no "14 points," no "Ten Commandments" in this book. This is a book of tools that work, not a book of rules.

The material in this book often seems like "common sense." Our work with many companies has shown that "sensible techniques" are more clearly understood, easier to correctly apply to a company's unique situation, and more widely accepted within the company—they work. Our sole measure of utility is whether a technique gets results in the real world.

Now, a bit of advice on reading the book. The chapters are designed to be self-contained. With the exception of Chapter 1, which explains why short cycle time is important, you will be able to obtain useful techniques by reading any single chapter. You will notice that the chapters interact positively with each other. As we will point out in the final chapter, Chapter 16, the whole can be larger than the sum of its parts.

Chapter 1 sets the stage, explaining when and why short development cycles are important. In our view, Chapter 2 is a "must-read" chapter because it describes the key tool of economic analysis. We cannot overemphasize the importance of this tool, since it quantifies the value of cycle time. By quantifying the alternatives, we can start making sound business decisions about cycle time. Chapter 3 describes the management of the "Fuzzy Front End." Chapters 4 through 6 concentrate on the product itself, illustrating how the product's concept can be shaped to facilitate rapid execution. Then chapters 7 through 13 systematically lay out the required management approach: staffing and organizing a team, managing overlapping activities, controlling progress, allocating resources, managing risk, and integrating manufacturing into development. Chapters 14 and 15 tie the subject together by explaining first how top management is involved in accelerating development projects and finally how to make the behavioral and attitudinal transition to rapid development. We have added a new chapter based on our experience during the last five years in working with managers to implement these tools. Two appendices explain how the approaches we are describing interact with the use of computerized development tools and with a variety of world-class manufacturing techniques.

This book will never be final, because we will continue to learn of improvements from those who are applying its tools in industry. The new content and more user-friendly format of this edition stem from what we see working in industry and the feedback of our readers. The next edition is likely to be heavily revised and include new examples. We sincerely welcome hearing from you about improvements you would like to see or contributions of examples from your experience.

We hope you find the book interesting, and more importantly, useful.

Preston G. Smith
New Product Dynamics
3493 NW Thurman St.
Portland, OR 97210
(503) 248-0900
INTERNET:preston@europa.com

Donald G. Reinertsen
Reinertsen & Associates
600 Via Monte D'Oro
Redondo Beach, CA 90277
(310) 373-5332
INTERNET:73770.2367@compuserve.com

Biographies

For over a decade, Preston G. Smith has concentrated exclusively on helping product development groups shorten their cycle times, initially as a staff consultant for a large diversified manufacturer and for the past nine years as an independent management consultant. He has developed and applied the techniques described in this book to products ranging from food and food packaging to medical electronics and motor vehicles. Prior to this specialization, he spent twenty years in engineering and management positions at IBM, AT&T, GM, and two small technology companies.

As a hands-on consultant, Preston has assisted dozens of companies in making the technical and cultural changes necessary to shorten cycle time. Thousands of product developers and managers have participated in his accelerated development workshops and seminars. He has taught accelerated product development at four universities and has published over a dozen articles on the topic.

He is a Certified Management Consultant and holds a Ph.D. in engineering from Stanford University.

Donald G. Reinertsen heads Reinertsen & Associates in Redondo Beach, California, a consulting firm specializing in the management of the product development process. He is recognized internationally for developing innovative analytical techniques for assessing product development processes. In 1983, while a consultant at McKinsey & Co., he wrote the classic article which first quantified the value of development speed. This article is often cited as the McKinsey study that indicated "Six months of delay can reduce a product's life cycle profits by 33 percent."

He has gone considerably beyond this early work. Drawing on extensive experience as both an operating manager and a management consultant, he consults with some of the world's leading companies on product development. He has trained hundreds of companies in a dozen countries in his seminars on rapid product development techniques.

Don writes and speaks frequently on techniques for shortening development cycles, and teaches a popular course at California Institute of Technology on Streamlining the Product Development Process. He has a B.S. in Electrical Engineering from Cornell, an M.B.A. from Harvard, and is a member of the IEEE, SME, and ASQC.

CHAPTER 1

The Time-to-Market Race

Manufacturing has become a much tougher, more global game during the 1970s and '80s, and many companies, particularly in North America, are not faring too well. Product innovation has come to be seen as a fundamental solution to these ills. But unfortunately the companies that could benefit most from having new products are often the ones that are unable to develop their products fast enough to keep up with turbulent, shifting markets. They wish they could bring products to market in half the time it now takes. This book guides the managers of these companies as they strive to quicken their new-products pace.

Increasing the pace to develop products in half the time is not a dream. Companies in many industries have made a reduction of this magnitude. Figure 1-1 shows a sampling of what companies in many different industries have been able to accomplish with specific products, working with both new and mature technologies. In fact, the figure suggests that our claim of a 50 percent reduction may be an understatement.

Moreover, the savings illustrated may be just the first step. Xerox, for instance, has cut its development cycles by 50 percent over the past decade, simultaneously making comparable improvements in product cost and quality, and plans to cut another year out of its cycles by 1993.

Many time lines for products, such as the ones shown in Figure 1-1, do not even include one of the biggest and most ignored opportunities to cut cycle lengths, one we call the fuzzy front end. This largely forgotten period, which runs from the time a company could have

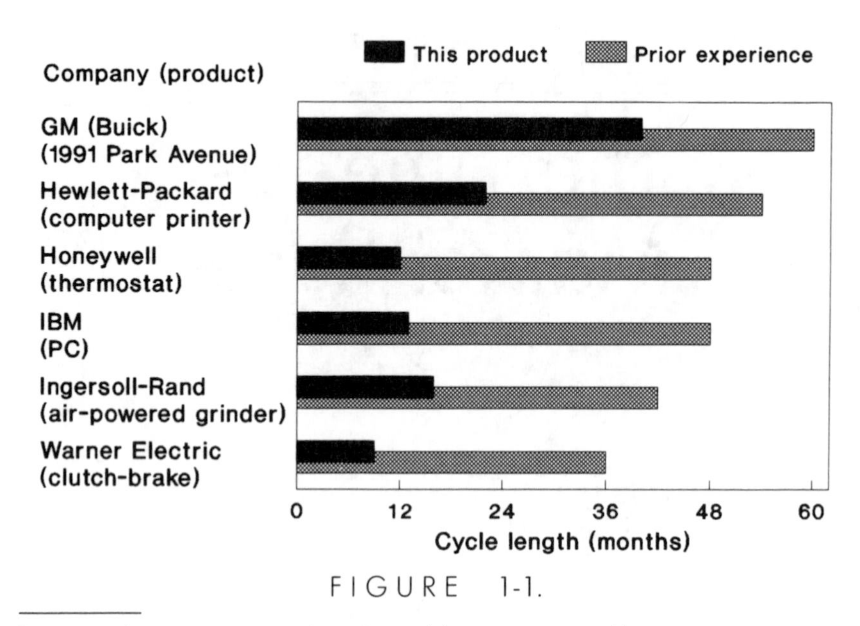

FIGURE 1-1.

In many industries, companies have been able to remove roughly half of the time formerly needed for product development.

started on a project until it actually does start in earnest, is the subject of Chapter 3.

Although a 50 percent reduction of the development cycle is a convenient reference, there is nothing magical about this number. Your firm can do either better or worse, depending on how advanced you are when you start measuring and how hard you work at it. We can aid you in making the transition more easily and effectively, but we cannot replace your own efforts. Companies that have done well at shortening development cycles have invested considerable effort in doing so. They are pleased with this investment because they have built a strategic competitive capability.

Fast product development is already an established practice in some industries but is just beginning to be recognized in others. The toy industry and many parts of the consumer electronics field, for instance, have become able to develop products quickly because competitive pressures have forced them to adopt approaches that work in a fast-changing marketplace. The personal computer industry, for one, moves quickly, because the underlying technology is advancing rapidly and is exploited by young, impatient companies like Apple and Compaq. As these industries demonstrate that increased development speed can be used to gain competitive advantage, other industries are likely to follow.

Leading companies around the globe are discovering that rapid product development is a huge, untapped source of competitive gain. Yet their existing management practices are generally not well suited to developing products quickly. Nowhere are the bureaucratic tendencies of companies more dangerous than in product development, an activity that attempts to integrate functions as disparate as marketing, sales, engineering, manufacturing, and finance, an activity that must transition from the chaos of invention to the daily discipline of the production environment.

Shortening the development cycle is a tool that no company can afford to ignore if it wants to remain viable in the 1990s. The challenge of making this transition is great, but so are the benefits of rapid product development.

THE BENEFITS OF A SHORT CYCLE

Many competitive advantages accrue from a fast development capability. Perhaps most obvious but least important is that the product's sales life is extended. If a product is introduced earlier, it seldom becomes obsolete any sooner. Consequently, for each month cut from a product's development cycle a month is added to its sales life, for an extra month of revenue and profit, as illustrated in Figure 1-2.

For some products that have high switching costs the benefit is even greater, because the early introducer gains on both ends of the cycle. If a product is introduced early, it gains more customers, who maintain their loyalty due to the cost of switching to another product. Their loyalty creates a sales tail that is roughly proportional to the prior sales of the product. Consequently, an earlier introduction develops momentum that not only carries the product's sales higher but also further into the future.

As a second benefit, early product introduction can increase market share. The first product to market has a 100 percent share of the market in the beginning. The earlier a product appears, the better are its prospects for obtaining and retaining a large share of the market. In some products like software and certain types of industrial machinery there is an especially high premium on offering the first product of a given type, because buyers get locked into the first operating system or computer language they acquire. It then becomes difficult for them to switch.

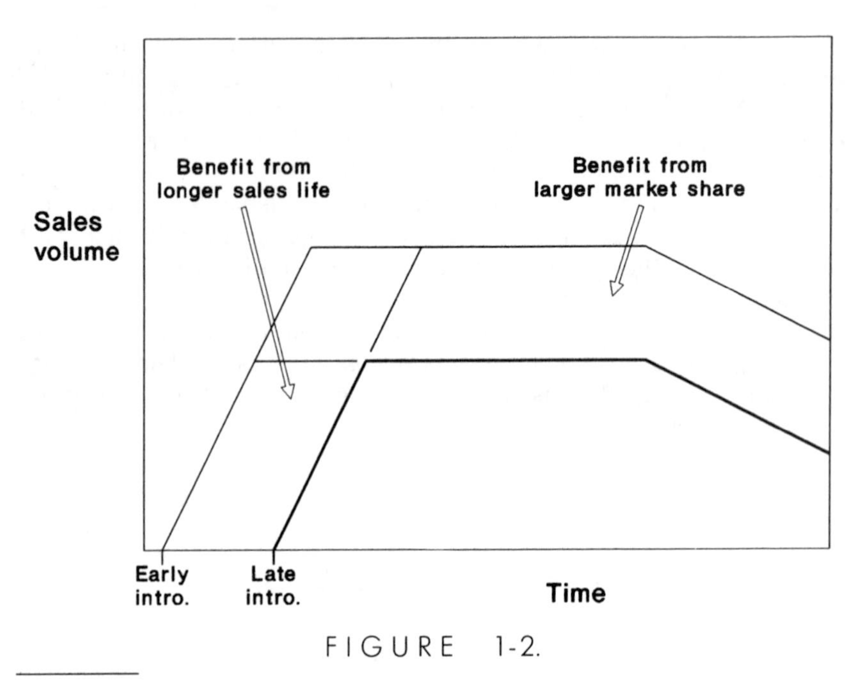

FIGURE 1-2.

Early introduction of a product can increase its sales life and market share.

A third benefit is higher profit margins. If a new product appears before there is competition, the company will enjoy more pricing freedom, making higher margins possible. The price may later decrease as competing products appear. By then, however, the company will be moving down the manufacturing learning curve ahead of the competition (see Fig. 1-3), so there is a continuing advantage to being first.

If a company is able to move quickly in new product development, it may occasionally choose to employ this capability by starting late rather than finishing early. There can be economic advantages to starting late when the underlying technology is moving down its price–performance curve rapidly. Today many products incorporate high-technology components such as microprocessors, whose price–performance characteristics improve rapidly. If an organization has the ability to develop products quickly, it can exploit technical opportunities by synchronizing its development with that of the latest technology. It then competes on product price or performance through a time-based strategy.

We once consulted for a company that was on the losing end of one of these situations. It had a durable old line of products that were entirely mechanical. The age of electronics was approaching, however,

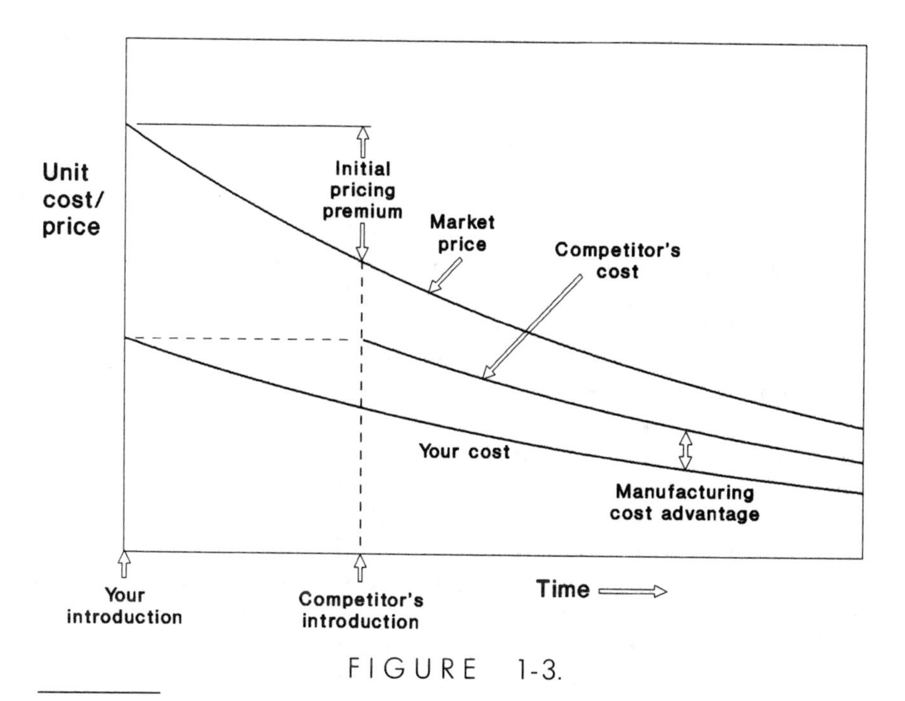

FIGURE 1-3.

Early entrants can enjoy premium pricing and cost advantages from the manufacturing learning curve.

so it started developing a microprocessor-based product. Unfortunately, one of its faster-moving competitors also decided to switch to microprocessor control. The competitor waited until a chip appeared that would do the job at half the cost, but then moved quickly. Although the two products appeared at about the same time, the competitor had a strong advantage, because its manufacturing cost was significantly lower.

An adaptable development process can be used to exploit changes other than technical ones. Many consumer markets change quickly, for unexpected reasons. Some products are sensitive to the vagaries of color, style, or current folk heroes. Political factors such as shifts in East–West relations often have an effect on many parts of the economy. The quick can turn these events into opportunities.

Finally, some of the benefits of fast product development are subtle and difficult to quantify but can nonetheless provide a great deal of competitive power. Some companies use speed as a strategy to create a perception of excellence continually. Companies like Apple, Honda, Matsushita, and 3M don't depend on somewhat greater profit margins or products' sales lives to create a competitive advantage from fast product development. Instead they focus on providing products that are viewed as fresh and innovative by end users and those in their distribution chain.

To gain advantage from rapid product development over the long haul a company must make swift, effective product development a way of life. In the end, the companies that can consistently develop new products quickly and effectively will step by step outpace those that cannot. Bringing one good product to market quickly is nice, but it will not ensure the viability of the company. Only the companies that have internalized this new way of operating will survive.

Industry managers know intuitively that faster product development often pays off directly as improved profitability for that product, but this is usually as far as their analyses go. Managers seldom calculate how much more they will actually make if a product is introduced a month earlier. Rather, they guess and debate the point, wasting more valuable time. In most cases it is possible to calculate the increased profitability resulting from changes in the development cycle's length, which helps managers make better business decisions and speeds up tactical decision making. Chapter 2 covers the techniques and benefits of quantifying the effects of faster development relative to other development objectives, such as lower cost and better product performance.

The benefits of rapid product development go beyond improving profitability of specific products and enhancing the company's image as an innovation leader. In fact, the attitudes and techniques that speed up the development process diffuse throughout the entire organization to improve the speed with which all resources, particularly the human one, can be applied. As companies grow they build greater resources for product development, but they also tend to become more bureaucratic, to cope with the problems of size. They often find that smaller, newer companies or ones in developing countries are able to beat them to market, even though these upstarts may be resource poor. Speeding up product development involves assessing how bureaucracy influences development speed, then making appropriate changes. Although this entire book relates to this topic, Chapter 10 (Monitoring and Controlling Progress), Chapter 12 (Managing Risk), and Chapter 14 (The Role of Top Management) deal with this subject extensively.

WHICH PRODUCTS JUSTIFY ACCELERATION?

Although the Manhattan Project was on a crash schedule, for good reason, we might feel more comfortable knowing that projects depending on nuclear fission are being conducted with a certain degree of deliberateness. How can we tell when a project should be on the fast track and when it merits a slower approach?

One technique is to make use of cost–benefit analysis to help make such decisions. There are often clear, quantifiable benefits available from rushing a project. We have already covered some of these, such as greater margins and longer sales life, and further discussion is provided in Chapter 2.

The costs of rushing are also quantifiable. They may include overtime charges, extra laboratory equipment, and similar development expenses. There could also be downstream consequences such as higher product costs or even greater product warranty expenses.

Once these factors have been put in some numerical perspective, it is possible to make a business decision whether it is desirable to rush a project to pursue a window of opportunity in the marketplace or whether the costs of doing so outweigh the potential gain. Chapter 2 provides the tools to make such decisions.

After comparing the costs and benefits, there is one other check to make. How large are the potential costs of acceleration, particularly the catastrophic costs such as product liability, relative to the assets of the company? Are we betting the company on accelerating one product?

Although accelerating anything can clearly increase the attendant risk, we should not leave the impression that fast product development is foolhardy. In fact, the market risk from going slowly may even be higher than the risk from proceeding quickly (see Chapter 12). Furthermore, some of the techniques we suggest for cutting time, such as using a tightly knit team to speed up decision making by making decisions at the working level (see Chapter 8), can actually enhance the quality of the design and thus reduce technical risk.

In some projects accelerated product development methods may be difficult to apply effectively. When dealing with large, complex projects, such as much aerospace and civil or power-plant construction work, the techniques presented here are not likely to fit well and may need adaptation. These projects seem to require complex organizational structure for adequate coordination and, because responsibility is so diffuse, complex control and reporting systems evolve to ensure that resources are channeled adequately and risk is managed with certainty. In contrast, work gets done much faster when the team can be kept small enough so that face-to-face communication and low-level decision making are the norm. With clever partitioning (see Chapter 6), some larger projects can be divided into fairly independent parts so that team techniques will still apply, but eventually a point is reached where it is no longer a team. This is the challenge the automotive companies face. Later chapters provide many examples of the creative approaches car makers are taking to shorten development cycles by employing small cross-functional teams.

The other area where fast development techniques may fail to provide major benefits is when basic technical discoveries are needed. Such projects fail to meet the guidelines for incremental innovation provided in Chapter 4. Because basic discoveries cannot be put on a timetable, forcing the process of invention is fruitless. Not only do new discoveries need some fermentation time, but the high-intensity scheduling that accompanies accelerated product development will burn people out if it is used on an extended project.

If you have the capability for fast development, it may be tempting to rush every product to market, but this temptation must be resisted. It often makes extra demands on resources and requires extra managerial attention to move a project through development quickly. It also takes committed, broadly skilled people, of which no company has a full complement, unfortunately. Rapid development is a valuable but special capability that is to be used when it will yield the greatest benefit. Chapter 10 describes how separate capabilities are needed for fast and regular projects, as is a means of assigning projects the proper track. Having distinctly different development processes in the same firm is uncommon, because most companies tend to run all projects through the same—usually slow—system, whether they fit or not.

It can be difficult to choose the few products that are to be accelerated from among many good candidates. This difficulty stems from the fact that most new products are in fact not so new, which tends to make them ideal candidates for rapid development. Theodore Levitt, in the 1966 article listed at the end of Chapter 4, observes that most of our new products are actually imitations and makes a strong case for being able to imitate quickly. Seldom does a truly new idea appear; products adapted from something else are much more prevalent. This being the case, why not build a product development system that can bring adaptations to market quickly? If we follow this logic, the first realization is that speed is not only desirable but imperative. We are clearly involved in a race against our competitors. The usual situation is that a competitor has either introduced or is about to introduce a product. You and probably other competitors would like to be involved in this market. You would like to enter it early, before there are many other competitors. It is necessary to enter it before the leader becomes too entrenched in the market and before it has proceeded very far down the manufacturing learning curve. Consequently, in this common scenario of a race between a leader and several other competing followers, you need to be the fastest among the followers. A process for dealing successfully with this situation should thus be part of your kit of basic business skills.

PUTTING THE TIME FACTOR INTO CONTEXT

One should not get the impression from our emphasis on time that other factors like product performance, quality, reliability, and cost are somehow less important. A company must be competitive in most, if not all, such areas in order to survive.

Before the 1980s, product cost and performance were the factors stressed most often. During the 1980s, quality became a focal point, forcing us to think in new and seemingly strange ways about how to run a business. During the 1990s, time to market is likely to be added to the requirements, which will also cause us to rethink our operations and change some habits. Just as the quality movement created some apparent conflicts and required training to enable workers to understand that, for example, quality and cost are not mutually exclusive, accelerating product development will require some reeducating too. Appendix B discusses the relationships between rapid product development and such programs as world-class manufacturing and total quality management.

There are good reasons for using time as the focal point to build a competitive product development process. The first, as we suggested at the outset, is that time has become an area of great opportunity for improvement. Technical advances such as those in computer graphics combined with stronger global competition have encouraged leading companies to search for ways of getting products to market much more rapidly. The companies that put effort into their development cycle are often able to shorten it by 50 percent. This forces their competitors to do likewise or be left behind with a catalog of products that have become stale in a few years.

Because development speed is supplementing rather than replacing other development objectives such as design for quality or for low manufacturing cost, it is becoming one more dimension to competitive advantage. In this regard we can learn from the Japanese, who cultivate the ability to learn one technique well, then absorb it and move on to the next without losing earlier gains. Relative to the emphasis placed on it in North America, the Japanese are now emphasizing product development speed, as Figure 1-4 suggests. The reason their current emphasis on product quality seems relatively low compared with that in North America is that they have for the most part mastered quality.

The other reason for concentrating here on development time is that it is a powerful concept on which to build. Time is easier for the general population to comprehend than terms like gross margin or market share. Everyone from the CEO to the shipping clerk understands

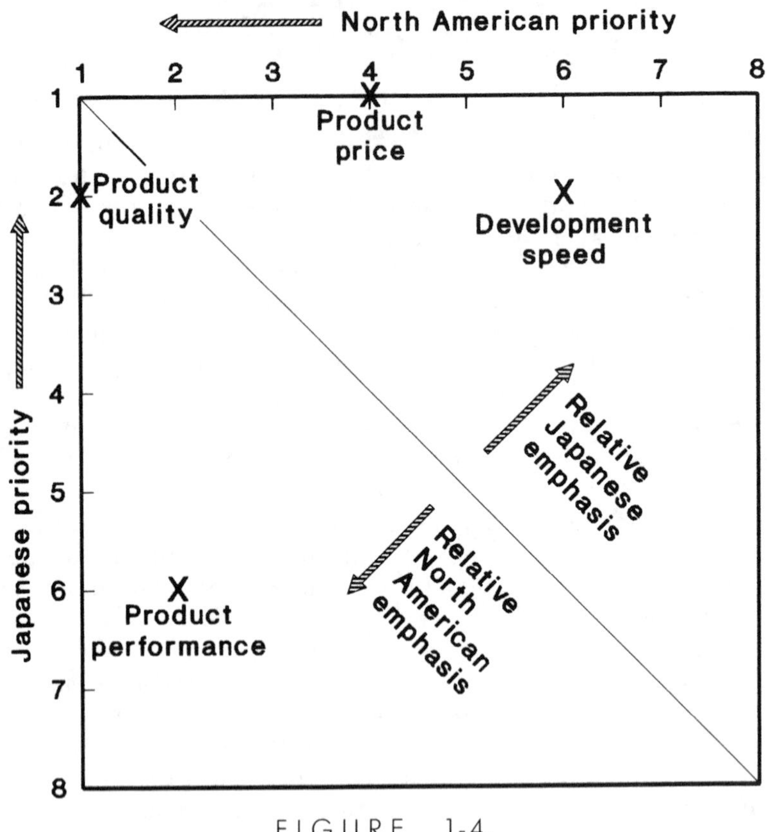

FIGURE 1-4.

Relative to North American conpanies, the Japanese are placing more emphasis on product development speed than on other development parameters. *From de Meyer et al.,* Flexibility: The Next Competitive Battle, *Boston University School of Management Manufacturing Roundtable, Table 4, p. 9, April 1987.*

how time slips away. Time has an intrinsic importance traceable to our own mortality.

Development time has direct competitive value, but it is also in a sense a surrogate for effectiveness of the product development process. "Effectiveness" tends to be a vague, uninteresting term, whereas "time" is easier to understand and measure.

Focusing on time can teach us about effectiveness in new and sometimes more graphic ways. Stressing time to market generally supports what managers are trying to achieve by getting the most out of their resources. The viewpoint of time is valuable because it often casts managerial choices in a new light, helping managers take steps that

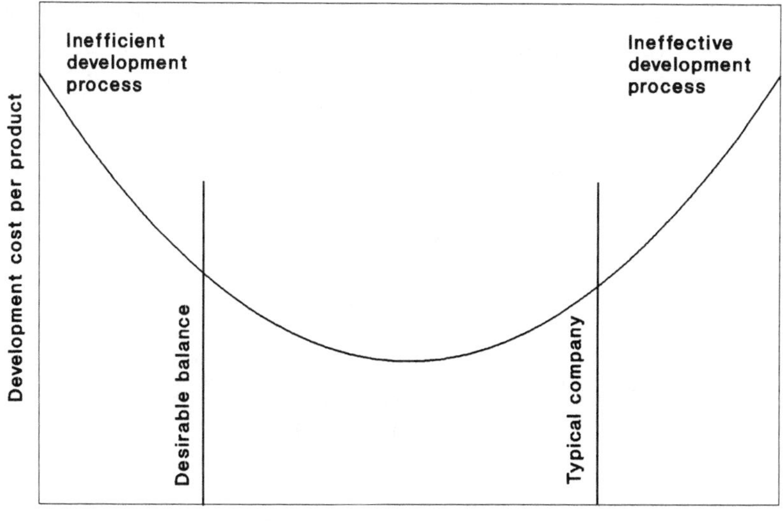

FIGURE 1-5.

It can be expensive to develop products too quickly, but most companies are far from this point.

might seem inadvisable from a traditional cost-effectiveness perspective. Consider an example. Managers are often reluctant to test prototypes by destroying them because handmade ones are expensive and their cost shows up directly in the budget. However, a test could save engineering labor needed otherwise for analysis. Time-based decision making might suggest using a destructive test. If cost effectiveness is then considered, it would be easier to see that engineering labor is more expensive than fabricating a prototype, even though the extra labor may not appear in the budget explicitly.

Sometimes managers fear it will be expensive to accelerate product development. They suspect that if the development cycle is compressed too much, resources will be used inefficiently. Our own experience suggests that most companies are far from this point, as indicated in Figure 1-5. When the acceleration techniques covered here are applied, the gains in effectiveness usually more than offset the losses from wasted motion. Professor Kim Clark of Harvard Business School has found, for example, that Japanese auto makers develop a car on average in 30 percent less calendar time and 50 percent less engineering hours than their North American counterparts, after adjusting for differences in vehicle size and complexity. Austin Rover of Great Britain has shortened car development time by 38 percent while also reducing development labor by 20 percent. Xerox reports that they have cut in half the resources and time required to develop comparable products. Honeywell has cut development time by 50 to 60 percent while decreasing labor hours 5 to 10 percent. Deere & Company has shrunk their seven-year development cycle by 60 percent while reducing development cost 30 percent. This subject and Figure 1-5 are explored in greater detail in Chapter 11.

SHIFTING FROM A COST TO A TIME MINDSET

Over the years, most businesses assimilate a plethora of accounting and control systems to measure costs with precision, and devote a great deal of effort to forecasting and monitoring sales data. But seldom is time analyzed or monitored with the same diligence. Few companies put as much effort on creating strategies for exploiting time as they spend on budgeting, cost reduction, or sales planning. Time thus remains a largely untapped source of competitive advantage.

To capitalize on this situation we must think of time and cost in much different terms. We must consider lost time as in fact a consumed resource and factor its cost into every decision, as we do with more conventional costs. The material in Chapter 2 on calculating the cost of time is therefore quite important.

Cost-oriented thinking frequently prevails over time-based thinking to the degree that often we are not even aware that we have a bias. In every aspect of product development examples abound of slow, uneconomical outcomes resulting from concentrating narrowly on cost. For instance, management commonly loads on plenty of projects to keep people busy, rather than realizing that the extra work creates queues and delays, which cut the profitability on new products. (Chapter 11 provides a quantitative explanation of the waste hidden in project overload.) Also, normal accounting methods tend to justify an equipment purchase based on how much it will be used (its cost per hour of operation) rather than on how much time it will save when it is needed. We justify first-aid kits based on their value in contingencies, but oscilloscopes have to be justified on their usage. More generally, accounting systems are set up to monitor costs precisely, but losses in schedules slip by routinely with no attempt made to determine how they affect the bottom line.

Changing from a cost-driven mindset to a time-driven frame will be most difficult for accountants, because most of their work involves taking time to monitor and control cost. Much of this work is essential to satisfying stockholders and tax collectors, but we must look constantly for ways to streamline accounting controls rather than setting them up as an obstacle course.

It is essential for nonfinancial people also to understand and be able to deal quantitatively with the financial value of time. Only when they are fluent in this language will they be able to challenge constructively cost systems that delay development.

As mentioned, the quality movement has forced a rethinking of some business concepts. The just-in-time (JIT) movement in manufacturing has done the same thing, which provides a particularly helpful analogy for us because it uses a time-based strategy. The goal of JIT manufacturing is not inventory reduction but being responsive to what the customer wants when it is wanted. JIT reduces factory lead time so that orders are filled faster, mistakes corrected sooner, forecasting becomes less critical, and a whole crew of production planning and expediting people formerly needed to cope with an essentially unpredictable market can be redeployed to add value. JIT does not concern itself with global optimizations such as factory utilization rates or economic order quantities. It is effective because it allows the factory to be more responsive to the marketplace.

JIT has received the benefit of much thought and application over the past decade, and its practitioners have countless success stories. There is much to be learned from it, both through direct analogy and by observing the change in viewpoint required to benefit from it. Subsequent chapters will transfer some JIT concepts to new product development.

BUILDING A HOLISTIC APPROACH

Our overall goal is to target dramatic reductions, not just minor improvements in time to market, which thus challenges much of the conventional wisdom on managing product development. Rapid product development is not a quick fix for getting one product—that is probably already late—to market faster. Instead it is a strategic capability that must be built from the ground up. The tough management issues emphasized include selecting people (Chapter 7), obtaining scarce resources (Chapter 11), integrating the R&D environment with production realities (Chapter 13), and the leadership of top management (Chapter 14).

In this book we strive to present a balanced, integrated view of rapid product development. This is not a marketing, engineering, or manufacturing book, though it goes into depth on practical issues in each of these areas. The greatest gains are to be made by integrating disciplines and considering their boundaries carefully.

The objective throughout is to explain just how product development can be accelerated—and how it can't. We avoid prescribing simplistic fixes like a renegade skunk works team to overcome problems overnight by evading bureaucracy. Instead we systematically dissect the product development process, pointing out what is needed at each stage, showing why certain techniques work, and emphasizing opportunities to shorten cycles. What results is an integrated management approach that allows critical products to be developed much faster, even in a corporate environment. This holistic view provides much more effective solutions than an engineering, marketing, or other functional viewpoint. This preference for an integrated view of rapid product development is apparent in the approach to writing specifications (Chapter 5) and for working on functional activities in parallel (Chapter 9).

Through applying these techniques in industry, we have learned that universal solutions are not effective, because each firm has different requirements and constraints. Furthermore, in order to create competitive advantage each solution must draw upon a company's strengths. Our approach is therefore one of probing the pros and cons of various approaches, respecting the complexity of the problem, and providing enough guidance for making final choices.

Although general techniques seem preferable, abstract solutions carry the danger of seeming idealistic and thus perhaps unusable. Consequently, numerous examples, both positive and negative, are applied to a broad range of products, to show how the various techniques apply. These applications have purposely been chosen to cover electronic and mechanical products, consumer and industrial goods, and both high- and low-volume production situations. Although the techniques presented normally apply across a range of company sizes from the small to the very large, we have purposely chosen most of our examples from among the more prominent companies, so that readers can appreciate the application in a familiar product and relate to it more easily.

Accelerated product development is primarily a distinctive management approach. Throughout, particularly in Chapters 14 and 15, we describe what this approach encompasses. At the outset, however, it may help to clarify our viewpoint by describing approaches that we believe to be too restrictive, although they are popular in the literature. One such view is to ascribe rapid development to computerized aids, such as computer-aided design (CAD); Appendix A covers the shortcomings of this view. Another type of shortsightedness is to equate accelerated product development with teamwork. There are in fact important elements of the solution that have more to do with the product than the team (see chapters 4–6), and the team itself is largely a mechanism to enable the use of other techniques such as overlapped activities (Chapter 9) and low-level decision making. Companies frequently stress teamwork and still don't bring products to market quickly.

Finally, many people connect accelerated product development with project management. Here again this perspective is too narrow, because it ignores the product-specific elements and the characteristics unique to product development projects. All too often "project management" fails because its key element is ignored: management does not provide the "project manager" with the authority or support necessary to reach the objective, an ambitious schedule in this case. We stress the importance of selecting and empowering the project manager. It is also easy to slow a project down by overuse of conventional project management techniques (see Chapter 10).

ARE WE HITTING THE MARK?

To the best of our knowledge this book was the first of its kind. Apart from some magazine articles, most of this material had never appeared before in print.

From what we have observed on-site in North American industry, industrial managers need this book today more than ever. We have therefore taken our own advice, from Chapter 4, and released this paperback edition even as we, the "developers," and you, the "customers," continue to learn.

Good products must always be guided by the needs of the customer, who is the best judge of these needs. We, like you, hope to provide improved products in the future, and thus solicit your help in this ongoing process. Please tell us about your successes and failures and where this book serves or fails to serve your needs, so that we can remain focused on the vital issues.

Preston G. Smith
New Product Dynamics
3493 NW Thurman St.
Portland, OR 97210
(503) 248-0900

Donald G. Reinertsen
Reinertsen & Associates
600 Via Monte D'Oro
Redondo Beach, CA 90277-6649
(310) 373-5332

SUGGESTED READING

Davis, Dwight B. 1989. Beating the clock. *Electronic Business* 15(11): 20–28.

Dumaine, Brian. 1989. How managers can succeed through speed. *Fortune* 19(4): 54–59. Two of several articles in the recent popular business press describing how companies are turning to time-based competition to gain advantage.

Stalk, Jr., George, and Thomas M. Hout. 1990. *Competing Against Time: How time-based competition is reshaping global markets.* New York: The Free Press, a division of Macmillan, Inc. A comprehensive account of how time is used to gain competitive advantage in business today, in product development as well as other areas, especially with respect to the Japanese threat.

Welter, Therese R. 1990. How to build and operate a product-design team. *Industry Week* 239(8): 35–58. A good example of recent articles focusing more narrowly on how companies are accelerating their development processes, although the emphasis in this piece is principally on teamwork, to the exclusion of the many other techniques we cover in this book.

CHAPTER 2

Wrapping It in Numbers

Until we attempt to evaluate the impact of rapid product development quantitatively, we are wandering in a vague world of emotional prejudices where the most articulate spokesperson wins out, regardless of the economic merit of his or her proposal. To avoid this we wrap numbers around our judgments.

Many companies are tempted to adopt a strategy of rapid product development solely because it feels like the right thing to do. As will be seen throughout this book, there are techniques for making these decisions on the basis of facts and analysis, and such an analytically based approach is likely to produce more consistently correct results. The point is not that everything can be precisely quantified but rather that enough useful information can be quantified to enable us to make systematic decisions. In the long run, analytically based decisions achieve better outcomes, with greater reliability, than the "gut feel" judgments made by most managers.

Over years of consulting experience we simply have not found that the average experienced manager has a good sixth sense of what the economics of development are in his or her business. Instead the majority of managers estimate their economics incorrectly, usually missing by hundreds of percentage points. The greatest errors appear to take place in underestimating the economic impact of slow development. Such miscalculations are important because they form the foundation upon which dozens of incorrect decisions may be based.

There are two very real benefits in quantifying the impact of speed. First, it helps management allocate scarce resources. The effort to accelerate product development is rarely an isolated program, but instead one of many initiatives competing for resources within a company. Only quantifying the impact of speed will allow intelligent decisions on what to spend in this effort and when to spend it. This quantification provides an objective way to communicate with other managers competing for scarce resources. Second, as will be seen later in this chapter, quantification provides some powerful rules of thumb. We need such rules to guide the hundreds of decisions that must be made in the course of the development program. These decisions involve trade-offs between speed and other objectives and must be made quickly and soundly. This can be done by providing good decision rules to guide the front-line workers developing the product.

We quantify the impact of speed by modeling the economics of a given development program and evaluating rapid product development as an objective in comparison with other possible objectives. The spreadsheet software that is most conveniently used to do this creates the impression of great accuracy, but do not be misled. Normally, managers do well to achieve an accuracy of two significant digits, but this is more than adequate to make the decisions confronting them. Furthermore, it is far better information than has been available in the past.

The resistance to quantification is sometimes high. Some people are tempted to avoid it because they feel it is difficult and imprecise. Yet it is much easier to quantify development programs than people who have never done so may suspect. In practice, such quantification is so powerful that even the crudest attempts to do it have produced extraordinary benefits for companies.

The remainder of this chapter explains how to quantify the importance of speed and to develop the decision rules needed to manage projects on a day-to-day basis.

THE FOUR KEY OBJECTIVES

In general it is useful to think of the task of managing the development of a new product as one of balancing efforts toward four key objectives (see Fig. 2-1). First we have development speed, or time to market. This is measured as the time between the first instant someone could have started working on a development program and the instant the final product was available to the customer. Consider the notion of a market clock that begins running when you or your competitors could actually

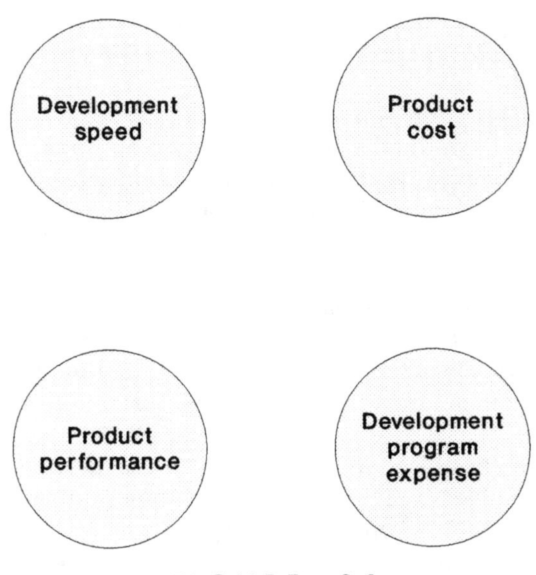

FIGURE 2-1.

Four key product development objectives.

have begun product development. There is no partial credit because someone in management sat on an idea for twelve months. If your competition was not also sitting on the idea, the clock was running. The market clock stops when the final product is available to the customer—in quantity. Simply producing artists' renderings or prototypes does not count. This is a more extended measure for the length of the development cycle than is recognized by most people, who would like to start the clock when management assigns a full-scale team to the program and stop it when the design is released to manufacturing. This narrow view arises from the tendency to view development as a task that takes place only in the engineering department. It fails to recognize that key development tasks and potential schedule delays also lie outside engineering.

The second key objective is product cost. Quite simply, this is the cost of the product delivered to the customer. It is important not to use merely the term "manufactured cost," because total cost will determine profits. A product with a low manufactured cost but a high sales and marketing cost may still be a loser. Cost includes both the one-time costs associated with manufacturing start-up and the recurring costs that appear in the cost-accounting system.

The third key objective is product performance—how well a product meets its market-based performance specification. A design that meets the needs of the marketplace has achieved good product performance. Product performance is how a product is rated in the eyes of the customer.

Performance, in the sense used here, is the same as product quality when quality is defined as conformance to customer requirements. We have chosen not to use the term "quality" because some organizations define it either more narrowly or broadly than others. When we refer to performance we require both a product that conforms to specification and a specification that fits market needs.

An excellent design must do more than just meet performance targets. It must achieve these targets using cost-effective technologies and design approaches. Often the majority of the time that goes into a design is spent on achieving a cost-effective solution rather than just a brute-force one.

The fourth key objective is development program expense. In relation to the other objectives this one is rarely important, although it may receive undue attention from managers under constant pressure to trim their budgets. Program expense includes all the one-time development costs associated with a specific project.

These four objectives are important simply because no one can pursue all four of them well at the same time. In some cases it may be possible to take actions that will help achieve more than one objective, but in other situations trade-offs must be made between objectives. In these cases management decisions should be made in a way that maximizes product profitability. Throughout this book we show many examples of situations where objectives must be balanced by using trade-off rules. To make wise trade-off decisions one must know the relative quantitative importance of the four key objectives.

THE SIX TRADE-OFFS

The art of managing product development depends on making good trade-offs between these four possible objectives. Developing an economic model for a development project creates the tool to make these trade-offs in a businesslike manner. Rather than simply taking the position that speed is important regardless of cost, managers can now have confidence in how far to go in pursuit of speed. They know whether $10,000 or $100,000 or $1,000,000 is the right amount to spend to speed up a schedule by one month.

To demonstrate how data from the economic model of a product help make decisions, let us illustrate the six trade-offs most likely to confront someone managing a development program (see Fig. 2-2). The product illustrated has expected sales of $100 million over a six-year product life. The product will average a 37 percent gross margin and produce an average 10 percent profit before tax. The incremental profit before tax, which is the extra profit earned for each incremental dollar of sales, will average 16 percent of sales.

Let us begin with the trade-offs related to speed. A common one is development speed versus product performance. For example, the development team may wish to add a feature to improve the product's performance, but it would take a little more development time to incorporate it. Knowing what the development time is worth and how much the extra performance is worth lets us make an intelligent business decision regarding this feature. If it would add two months of delay that are valued at $470,000 per month, the cost of the delay is $940,000. The extra feature must therefore add at least this much to the product's life-cycle profits. If we assume that the feature adds 1 percent to unit sales, it will create an extra $1 million of revenue over the life cycle of the product. If 16 percent of the incremental sales dollar is converted into profit before tax, the feature would be worth only $160,000 in profit impact. In this case it would be a poor decision to delay the product to add this feature (see Fig. 2-3). In fact, it is quite likely that the feature could be added during a model extension, or else saved for a future product (see Chapter 4).

In the same way, we can make decisions trading off development speed against product cost. Assume that a product has a subassembly that we plan to reengineer. We discover that the new technology we had planned to use will now not be available from the supplier for another six months. With the old technology, the cost of the subassembly will be 20 percent higher than planned. If the subassembly represents 10 percent of the cost of the product, the old technology would raise costs by 2 percent. Suppose further that this added cost of using the old technology would appear only in units accounting for the first 10 percent of revenue, and during this period we planned to have an average gross margin of 50 percent. This could occur if we planned to switch to the new technology at about the time that the first 10 percent of the units were sold. In a case like this, again assuming a value of $470,000 per month lost, the six months of delay would be worth $2,800,000. Only $10 million of the life cycle sales of $100 million would be affected by the higher costs, and these sales would have a 1 percent reduction in gross margin. In total, this would reduce life cycle profits by $100,000 (see Fig. 2-4). Based on these facts, it is a sound business decision to produce the first 10 percent of the products with

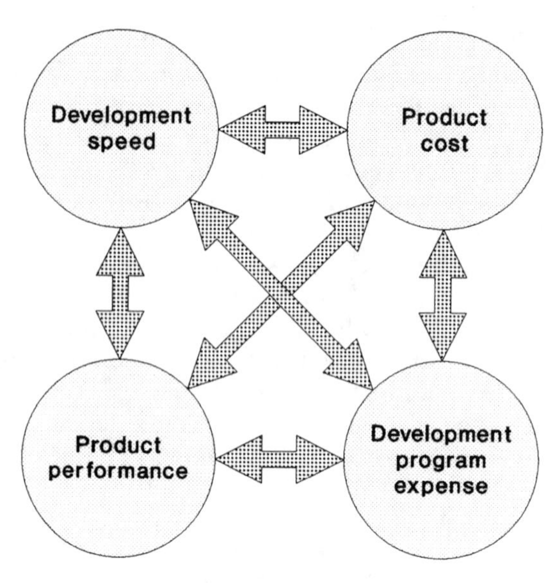

FIGURE 2-2.

The six trade-offs between product development objectives.

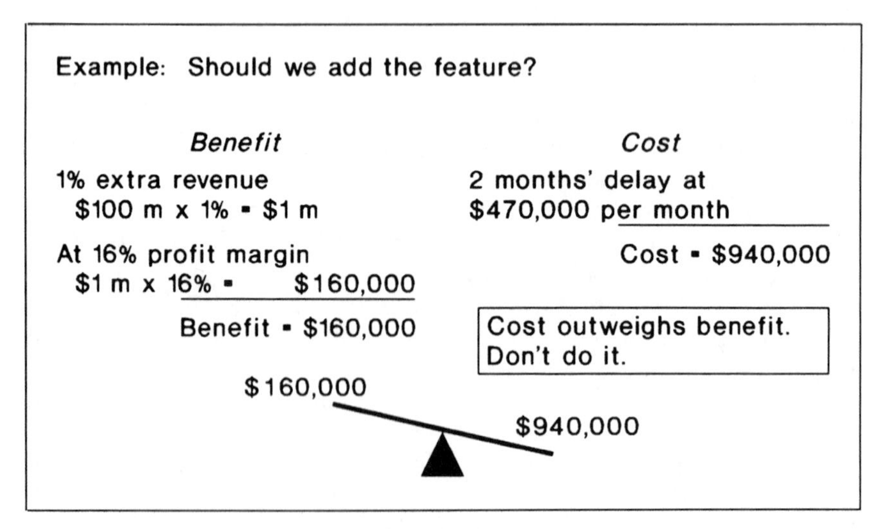

FIGURE 2-3.

The trade-off of product performance versus development speed.

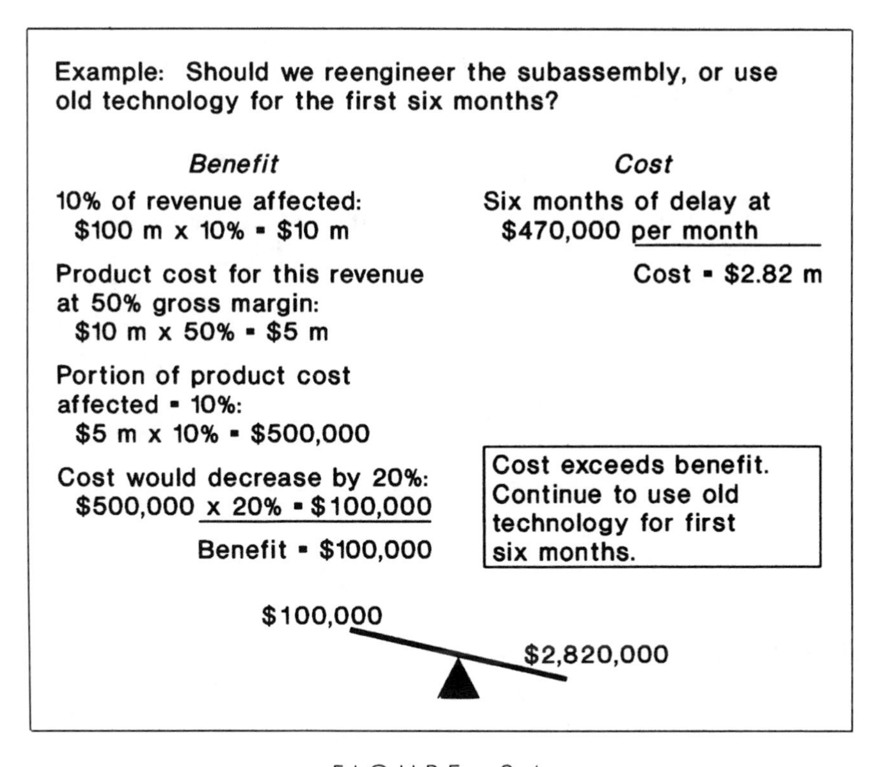

FIGURE 2-4.

The trade-off of product cost versus development speed.

the old technology in the affected subassembly, then introduce the new technology as a running change. Doing so would produce a net savings for the company of $2,700,000.

Here is another common trade-off. Suppose that an outside drafting service is being considered to try to speed up the preparation of detail drawings on this project. In this case two weeks can be cut from the schedule by using the outside service to do 200 hours of drafting at $25 per hour. If the schedule is worth $470,000 per month, two weeks are worth about $240,000. To spend $5,000 on the outside drafting would thus be an attractive business decision (see Fig. 2-5).

Another common trade-off is the one between performance and product cost. For example, marketing may dream up a feature that they believe will add 6 percent to the unit sales volume over the life of the product. Engineering then estimates that this feature will increase the product cost by 10 percent. If our average gross margin over the life of the product is 37 percent, the cost increase would be $6.3 million on

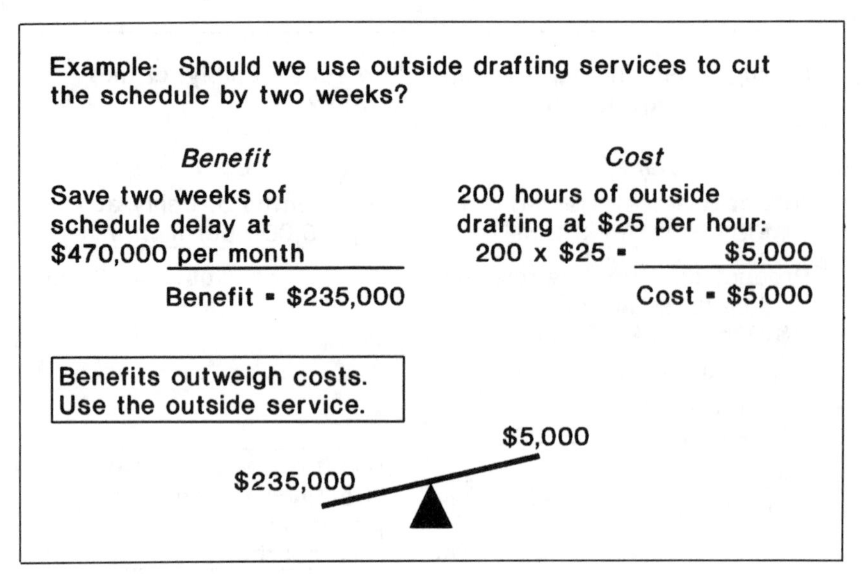

FIGURE 2-5.

The trade-off of development expense versus development speed.

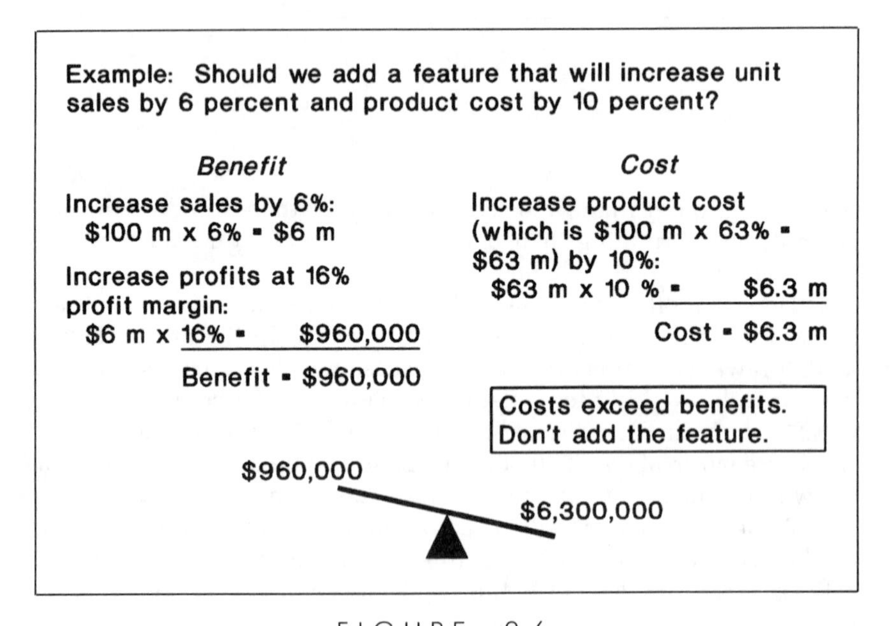

FIGURE 2-6.

The trade-off of product performance versus product cost.

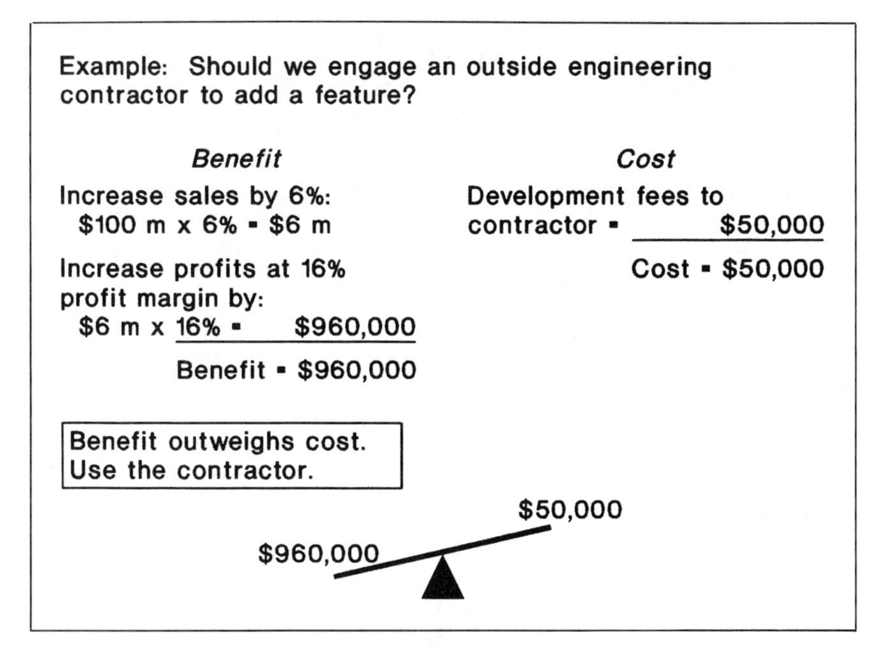

FIGURE 2-7.

The trade-off of product performance versus development expense.

a product generating $100 million in sales. The extra $6 million of sales volume will produce a profit of $960,000, assuming the same 16 percent incremental profit margin discussed earlier. It would clearly be unattractive to add this much cost to the product to chase after such a small profit potential (see Fig. 2-6).

Similarly, we may find ourselves trading off product performance against development expense. Perhaps an outside engineering contractor has agreed to add a feature to the product without taking time from its schedule. To do this will increase development expense by $50,000. If the increased performance would increase sales by 6 percent and 16 percent of this became pretax profits, the $50,000 investment in extra development expense would produce $6 million in incremental sales and $960,000 in incremental profits. It would therefore be a good idea to use the outside contractor (see Fig. 2-7).

Finally, we may find ourselves trading off product cost against development expense. For example, let's say that by spending an extra $50,000 we can reengineer a certain subassembly to remove 15 percent of its costs. The costs account for 10 percent of overall product cost, and we anticipate a 37 percent gross margin. In this case we have an

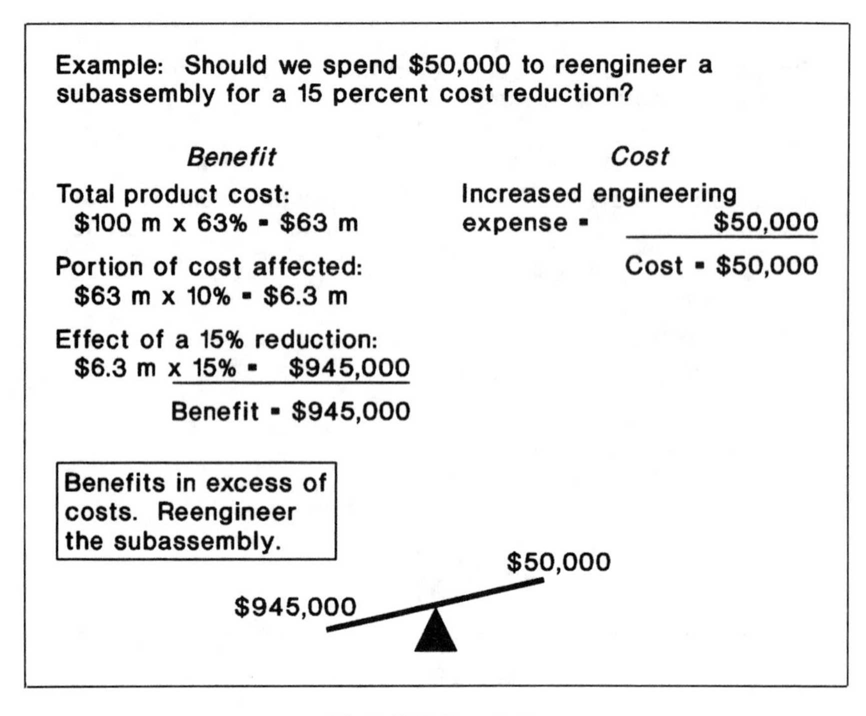

FIGURE 2-8.

The trade-off of product cost versus development expense.

opportunity to improve profits by 0.9 percent. If life cycle profits are $100 million, we should get a return of about $900,000 on this effort, which would be a very attractive return (see Fig. 2-8).

We have just illustrated the six types of trade-offs that must be made daily in the course of a development project. The power of the economic model is that it tells where to draw the line. What may be a good investment at $50,000 could turn out to be a bad one at $150,000. With no financial yardsticks, it is impossible to decide what to do.

Clearly, a good economic model can handle even more complex decisions. For instance, the opportunity could arise to spend an extra $50,000 on development, raise the product's cost by 0.5 percent, and add four weeks to the development schedule in return for a 15 percent increase in product sales. Using the same model seen earlier, the cost impact would include $50,000 of development expense, $315,000 of costs related to product cost, and $470,000 of schedule-related costs, for a total cost of $835,000. In return we would expect a $15 million

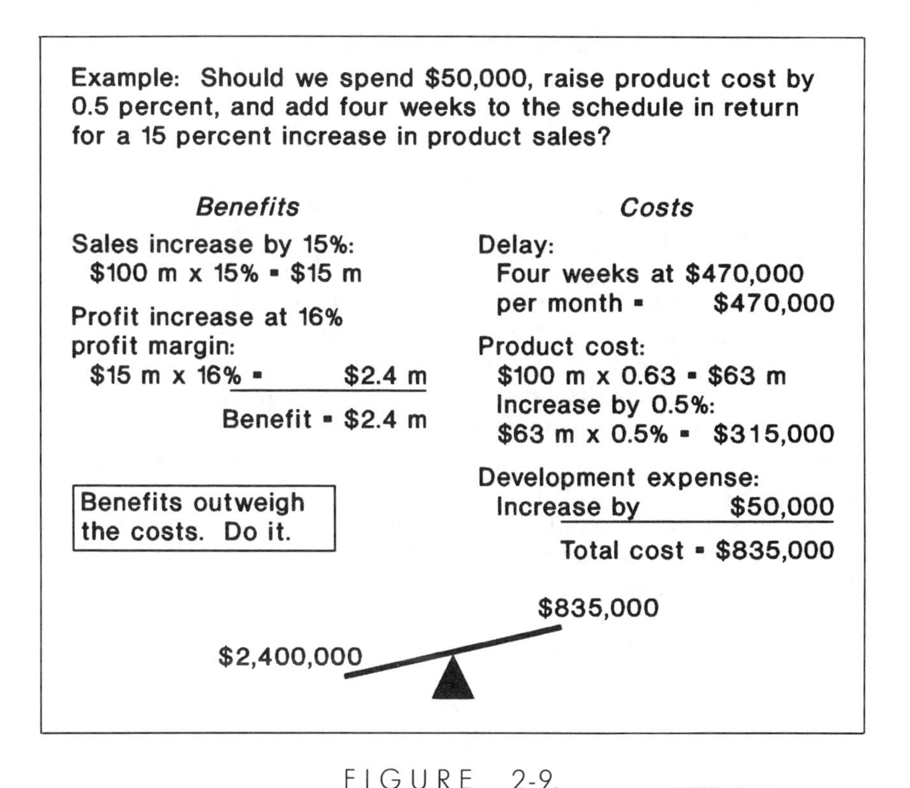

FIGURE 2-9.

A complex, real world, development decision.

increase in sales and a $2.4 million increase in profits. In this case we should take advantage of the opportunity (see Fig. 2-9).

These are not hypothetical decisions—in a real development project, similar decisions take place every day. They illustrate the value of having an effective model for the economics of a development program. Simply put, no one can make these decisions well by relying exclusively upon intuition. Furthermore, even the extraordinary manager who might make them well will have subordinates unlikely to have that level of skill. With good decision rules, these subordinates can decide quickly and correctly. If they do not have them, progress will stop until the manager can be located to make the decision.

How hard is it to prepare such a model? Does it take a team of ten MBAs and mainframe computer? Fortunately, it is much simpler than most people suspect. The next section discusses how such a model is prepared.

CREATING THE MODEL

Economic models for development projects can be constructed in great detail or be done very quickly. Although we have at times used such sophisticated techniques as net present value, discounted cash flow analysis, and internal rate of return analysis, we have not found these techniques to be superior to a simple comparison of cumulative profits over the life of the product under different cases. This is particularly true for products with sales lives of five years or less.

A simple model is adequate because it is likely to be far more accurate than some of the data used as input. The outcome of the analysis will depend heavily on certain critical assumptions such as projected average selling price and unit sales levels. These particular assumptions have an intrinsic degree of uncertainty that neutralizes the value of highly sophisticated financial analysis techniques. This uncertainty prevents taking advantage of these techniques' sophistication. Furthermore, this sophistication leads to two disadvantages. First and most important is that it makes it extremely difficult to communicate results to the non-MBA. Success is achieved by enabling people to understand our conclusions and react to them. Any obstacle in this communication process should be avoided if possible. Second, using sophisticated techniques may create the impression of greater precision than is actually being achieved. The sad story of most complex, black-box models is that they are used and trusted by organizations long after they have ceased to model reality well. Their very complexity prevents the organization from realizing that their trust in the model is misplaced. The real world requires simple models that are easily understood. Particularly when input data are imprecise, we should be more than willing to sacrifice precision for ease of communication.

The most useful technique we have found is to build a baseline model and several equally probable variations on it. The variations should be chosen as roughly equally probable shortfalls in development objectives. For example, they might be overrunning the development budget by 50 percent, introducing the product six months late, missing the product-cost target by 10 percent, or having performance shortcomings that reduce unit sales by 10 percent (see Fig. 2-10). Starting with the baseline case we calculate the life-cycle profits of each case, then compare the differences between cases. These differences are finally used to develop trade-off rules.

Profit impact should always be calculated over the full expected life of the product. It should start at the beginning of the development project before any revenue is produced and continue until the product

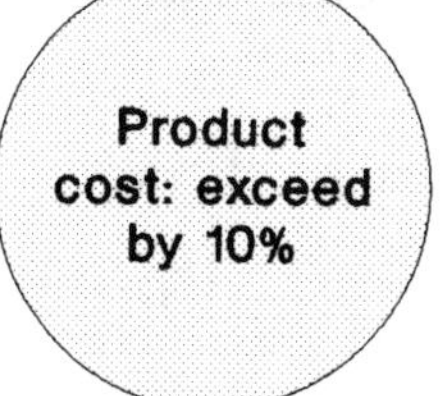

Product performance: lose 10% of unit sales (due to product shortcomings)

Development program expense: exceed by 50%

FIGURE 2-10.

Equally likely shortfalls in development objectives.

is eventually dropped. Figure 2-11 shows a typical profit projection over the six-year life of a product.

The projections are best developed by a team including representatives from engineering, marketing, manufacturing, and finance. In most cases the financial analyst is likely to be a good choice to build and maintain the model. He or she is likely to be reasonably objective and will probably be pleasantly surprised at how simple a financial model we are using.

The next few paragraphs describe what is in the model in Table 2-1, which it may be helpful to refer to as we proceed. First, project the average sales price of a product over its expected life. These prices typically fall at a rate proportional to the rate of improvement in price–performance of a product's underlying technology. For example, semiconductor unit prices typically fall at a rate of 25 to 30 percent per year. Computer printers may decrease by 15 percent per year. Disk drives characteristically drop at about 20 percent per year, mechanical systems at about 5 percent. Each market normally has an underlying long-term pricing trend that is determined by price–performance changes in its underlying technologies. In the table we use a 10 percent rate of

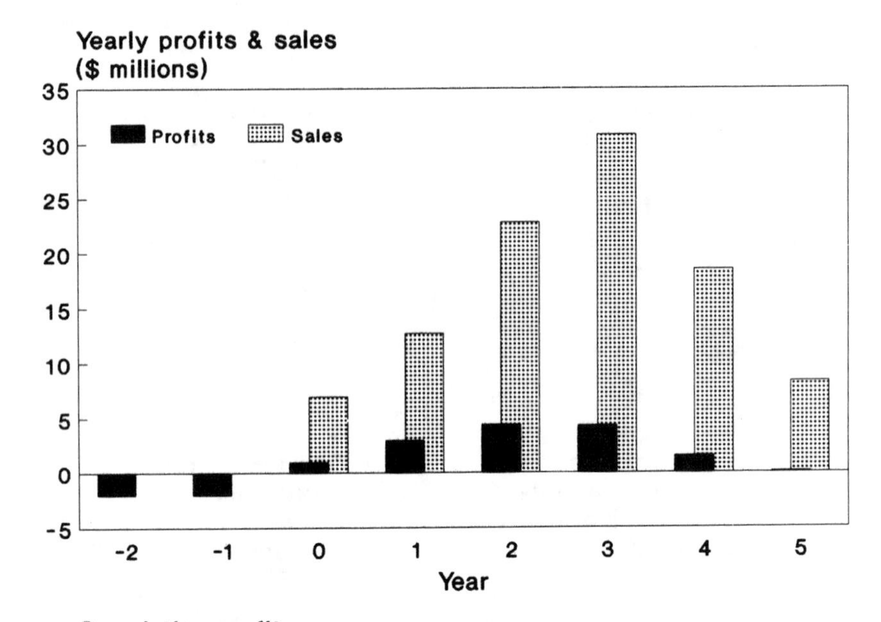

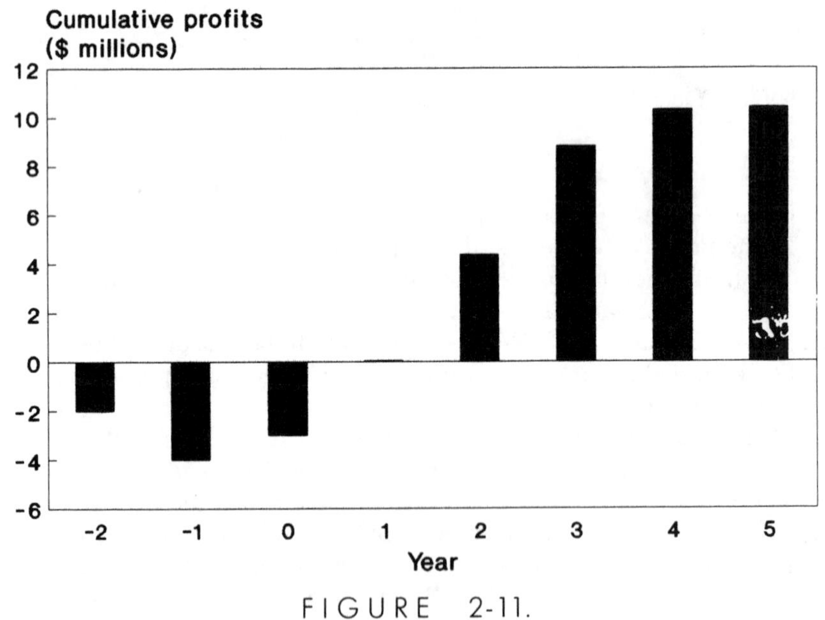

FIGURE 2-11.

Yearly and cumulative results from a baseline model.

TABLE 2-1

A Baseline Product Profit Model.

		Year							
	Assumes:	−2	−1	0	1	2	3	4	5
Average sales price	−10%/year			$7,033	$6,330	$5,697	$5,127	$4,614	$4,153
Introduction price	$7,033								
Market size (units)				10,000	20,000	40,000	60,000	40,000	20,000
Market share				10.0%	10.0%	10.0%	10.0%	10.0%	10.0%
Unit sales				1,000	2,000	4,000	6,000	4,000	2,000
Dollar sales				$7,033,000	$12,659,400	$22,786,920	$30,762,342	$18,457,405	$8,305,832
Unit Cost	−2%/year			$3,516	$3,446	$3,377	$3,309	$3,243	$3,178
Introduction cost	$3,516								
Cost of goods sold				$3,516,000	$6,891,360	$13,507,066	$19,855,386	$12,972,186	$6,356,371
Gross margin dollars				$3,517,000	$5,768,040	$9,279,854	$10,906,956	$5,485,219	$1,949,461
Gross margin percent				50.0%	45.6%	40.7%	35.5%	29.7%	23.5%
Engineering		$2,000,000	$2,000,000	$1,000,000	$100,000	$100,000	$100,000	$100,000	$100,000
Marketing	16% sales			$1,125,280	$2,025,504	$3,645,907	$4,921,975	$2,953,185	$1,328,933
G&A	5% sales			$351,650	$632,970	$1,139,346	$1,538,117	$922,870	$415,292
Operating expense		$2,000,000	$2,000,000	$2,476,930	$2,758,474	$4,885,253	$6,560,092	$3,976,055	$1,844,225
Profit before tax (PBT)		($2,000,000)	($2,000,000)	$1,040,070	$3,009,566	$4,394,601	$4,346,864	$1,509,164	$105,237
Cumulative PBT		($2,000,000)	($4,000,000)	($2,959,930)	$49,636	$4,444,237	$8,791,101	$10,300,265	$10,405,502
Return on sales (PBT/$ sales)				14.8%	23.8%	19.3%	14.1%	8.2%	1.3%
Cumulative sales				$100,004,900					
Cumulative gross margin				$36,906,531					
Cumulative PBT				$10,405,502	Baseline				
Average percent gross margin				36.9%					
Average percent return on sales				10.4%					

annual decrease. This drop in pricing eventually causes the manufacturer to discontinue the product because it can no longer be produced profitably.

After estimating average sales prices over the life of the product unit sales are projected. Unit sales are a function of both market size and market share. Typically, sales will grow exponentially until they reach a peak, then slowly trail off. A common exception to this type of behavior is OEM (original equipment manufacturer) sales, which may remain low for a longer period of time in the beginning but rise rapidly once customer shipments start. The nature of the selling cycle determines this pattern. In OEM sales we often place demonstration units in the hands of the OEMs so that they can evaluate them and design them into their systems. Once the products are designed in, sales may rise rapidly as the OEM begins to ship the system in quantity.

After projecting dollar sales the table projects the product's unit costs, which usually decrease because of the learning curves common in the industry. These costs almost always go down at a slower rate than prices drop. Typical learning curves might produce cost drops of 5 to 15 percent for every doubling of volume. The latter value would be called an 85 percent learning curve. The most sophisticated cost models will establish a combination of price trends and learning curves for material, labor, and overhead components of product cost. Manufacturing should be involved in projecting these costs, since they will be familiar with product cost behavior. For simplicity the model simply shows unit costs dropping at 2 percent per year.

The factors mentioned so far allow us to calculate gross margin over the life of the product. It can be seen to drop until eventually the product must be discontinued.

The next step is to project operating expenses. In addition to the expenses shown in Table 2-1, one-time expenses such as tooling can be shown as a separate line item. Engineering expenses are direct estimates associated with the project. Marketing expenses may include one-time estimates associated with the specific program, as well as a portion of the selling and marketing overhead applied as a straight percentage of sales. General and administrative expenses are normally applied as a straight percentage of sales. The percentages to use in each case are simply those that are customary for the business in question.

After all expenses have been included, calculate the profit before tax (PBT). We find it most useful to focus on cumulative profit before tax as a simple, easily understood measure of the development program. In our experience more-complicated measures add no additional insight but simply complicate communications. The numbers behind the baseline case are shown in Table 2-1.

Once the baseline case has been completed, go to the same people and have them estimate values for the variations. By using the same people we ensure that the same biases are built into the assumptions for each variation. The adjustments for development expenses overruns are obvious (see Table 2-2).

When computing cases where the product fails to reach its cost targets, recognize that it is not necessary to assume that this cost problem is borne throughout the life of the product. Rather in most cases it is realistic to assume that the product will be reengineered after a year or two of sales, to restore margins to the level that would have been reached if the original product cost target had been achieved (see Table 2-3).

Performance problems are normally approximated by having unit sales drop by a fixed percentage, although at times an acceptable alternative is to discount the average sales price, since this is often done to retain market share with a slightly inferior product. It is important to be careful in using such price discounting because it has severe effects on life-cycle profits (see Table 2-4).

The most complex case to model is a delay in product introduction. Generally most products fit into one of the following three cases. First, there are inelastic products with limited competition. For these products, total unit sales do not decrease. Instead, pent-up demand will cause the product's cumulative unit sales to remain constant even if it is introduced late. This can occur when loyal customers are locked in to the product and relatively indifferent about its introduction date. In this case the cost of schedule delay comes simply from price erosion because units are shipped during a relatively low-margin period in the product's life. This delay is generally the least severe of the possible schedule delays (Fig. 2-12).

In the second case there is a somewhat smaller sales peak. Sales build at about the same rate as they would have with an earlier introduction, and the peak occurs in the same year as it would have under the baseline scenario. In this case market demand for the product is determined by external forces, so that it peaks at the same time regardless of when the product is introduced. There is simply a somewhat lower market share throughout the life of the product when compared to the baseline case. This causes a more severe decrease in profits, but not as much as in the third and final case.

In the third case we see a severe, continuing reduction in sales. This decline can occur in products such as medical devices, where a six-month introduction delay may mean the difference between a 20 percent market share and a 2 percent share. The same effect occurs in many OEM markets where early introduction often results in design-in victories and the early winner cannot be dislodged by later competitors.

In preparing a model, pay the most attention to unit sales assump-

TABLE 2-2

A Product Profit Model When Development Expense Overruns by 50 Percent.

		Year							
	Assumes:	−2	−1	0	1	2	3	4	5
Average sales price	−10%/year			$7,033	$6,330	$5,697	$5,127	$4,614	$4,153
Introduction price	$7,033								
Market size (units)				10,000	20,000	40,000	60,000	40,000	20,000
Market share				10.0%	10.0%	10.0%	10.0%	10.0%	10.0%
Unit sales				1,000	2,000	4,000	6,000	4,000	2,000
Dollar sales				$7,033,000	$12,659,400	$22,786,920	$30,762,342	$18,457,405	$8,305,832
Unit Cost	−2%/year			$3,516	$3,446	$3,377	$3,309	$3,243	$3,178
Introduction cost	$3,516								
Cost of goods sold				$3,516,000	$6,891,360	$13,507,066	$19,855,386	$12,972,186	$6,356,371
Gross margin dollars				$3,517,000	$5,768,040	$9,279,854	$10,906,956	$5,485,219	$1,949,461
Gross margin percent				50.0%	45.6%	40.7%	35.5%	29.7%	23.5%
		Higher development expense							
Engineering		$3,000,000	$3,000,000	$1,500,000	$150,000	$150,000	$150,000	$150,000	$150,000
Marketing	16% sales			$1,125,280	$2,025,504	$3,645,907	$4,921,975	$2,953,185	$1,328,933
G&A	5% sales			$351,650	$632,970	$1,139,346	$1,538,117	$922,870	$415,292
Operating expense		$3,000,000	$3,000,000	$2,976,930	$2,808,474	$4,935,253	$6,610,092	$4,026,055	$1,894,225
Profit before tax (PBT)		($3,000,000)	($3,000,000)	$540,070	$2,959,566	$4,344,601	$4,296,864	$1,459,164	$55,237
Cumulative PBT		($3,000,000)	($6,000,000)	($5,459,930)	**($2,500,364)**	$1,844,237	$6,141,101	$7,600,265	$7,655,502
Return on sales (PBT/$ sales)				7.7%	23.4%	19.1%	14.0%	7.9%	0.7%
Cumulative sales				$100,004,900					
Cumulative gross margin				$36,906,531					
Cumulative PBT				$7,655,502	Lower profit				
Average percent gross margin				36.9%					
Average percent return on sales				7.7%					

TABLE 2-3

A Product Profit Model When the Product Cost Overruns by 10 Percent for Two Years.

	Assumes:	−2	−1	0	1	2	3	4	5
		Year							
Average sales price	−10%/year			$7,033	$6,330	$5,697	$5,127	$4,614	$4,153
Introduction price	$7,033								
Market size (units)				10,000	20,000	40,000	60,000	40,000	20,000
Market share				10.0%	10.0%	10.0%	10.0%	10.0%	10.0%
Unit sales				1,000	2,000	4,000	6,000	4,000	2,000
Dollar sales				$7,033,000	$12,659,400	$22,786,920	$30,762,342	$18,457,405	$8,305,832
	−2%/year		Two years of higher unit cost	$3,868	$3,790	$3,377	$3,309	$3,243	$3,178
Unit Cost									
Introduction cost	$3,868								
Cost of goods sold				$3,867,600	$7,580,496	$13,508,000	$19,856,760	$12,973,083	$6,356,811
Gross margin dollars				$3,165,400	$5,078,904	$9,278,920	$10,905,582	$5,484,322	$1,949,022
Gross margin percent				45.0%	40.1%	40.7%	35.5%	29.7%	23.5%
Engineering		$2,000,000	$2,000,000	$1,000,000	$100,000	$100,000	$100,000	$100,000	$100,000
Marketing	16% sales			$1,125,280	$2,025,504	$3,645,907	$4,921,975	$2,953,185	$1,328,933
G&A	5% sales			$351,650	$632,970	$1,139,346	$1,538,117	$922,870	$415,292
Operating expense		$2,000,000	$2,000,000	$2,476,930	$2,758,474	$4,885,253	$6,560,092	$3,976,055	$1,844,225
Profit before tax (PBT)		($2,000,000)	($2,000,000)	$688,470	$2,320,430	$4,393,667	$4,345,490	$1,508,267	$104,797
Cumulative PBT		($2,000,000)	($4,000,000)	($3,311,530)	($991,100)	$3,402,567	$7,748,057	$9,256,324	$9,361,121
Return on sales (PBT/$ sales)				9.8%	18.3%	19.3%	14.1%	8.2%	1.3%
Cumulative sales				$100,004,900					
Cumulative gross margin				$35,862,150					
Cumulative PBT				$9,361,121	Lower profit				
Average percent gross margin				35.9%					
Average percent return on sales				9.4%					

TABLE 2-4

A Product Profit Model When Product Performance Problems Reduce Unit Sales by 10 Percent.

	Assumes:	Year −2	−1	0	1	2	3	4	5
Average sales price	−10%/year			$7,033	$6,330	$5,697	$5,127	$4,614	$4,153
Introduction price	$7,033								
Market size (units)			Unit sales	10,000	20,000	40,000	60,000	40,000	20,000
Market share			down by	9.0%	9.0%	9.0%	9.0%	9.0%	9.0%
Unit sales			10 percent	900	1,800	3,600	5,400	3,600	1,800
Dollar sales				$6,329,700	$11,393,460	$20,508,228	$27,686,108	$16,611,665	$7,475,249
Unit Cost	−2%/year			$3,516	$3,446	$3,377	$3,309	$3,243	$3,178
Introduction cost	$3,516								
Cost of goods sold				$3,164,400	$6,202,224	$12,156,359	$17,869,848	$11,674,967	$5,720,734
Gross margin dollars				$3,165,300	$5,191,236	$8,351,869	$9,816,260	$4,936,697	$1,754,515
Gross margin percent				50.0%	45.6%	40.7%	35.5%	29.7%	23.5%
Engineering		$2,000,000	$2,000,000	$1,000,000	$100,000	$100,000	$100,000	$100,000	$100,000
Marketing	16% sales			$1,012,752	$1,822,954	$3,281,316	$4,429,777	$2,657,866	$1,196,040
G&A	5% sales			$316,485	$569,673	$1,025,411	$1,384,305	$830,583	$373,762
Operating expense		$2,000,000	$2,000,000	$2,329,237	$2,492,627	$4,406,728	$5,914,083	$3,588,450	$1,669,802
Profit before tax (PBT)		($2,000,000)	($2,000,000)	$836,063	$2,698,609	$3,945,141	$3,902,177	$1,348,248	$84,713
Cumulative PBT		($2,000,000)	($4,000,000)	($3,163,937)	$465,328)	$3,479,813	$7,381,991	$8,730,239	$8,814,952
Return on sales (PBT/$ sales)				13.2%	23.7%	19.2%	14.1%	8.1%	1.1%
Cumulative sales				$90,004,410					
Cumulative gross margin				$33,215,878					
Cumulative PBT				$8,814,952	Lower profit				
Average percent gross margin				36.9%					
Average percent return on sales				9.8%					

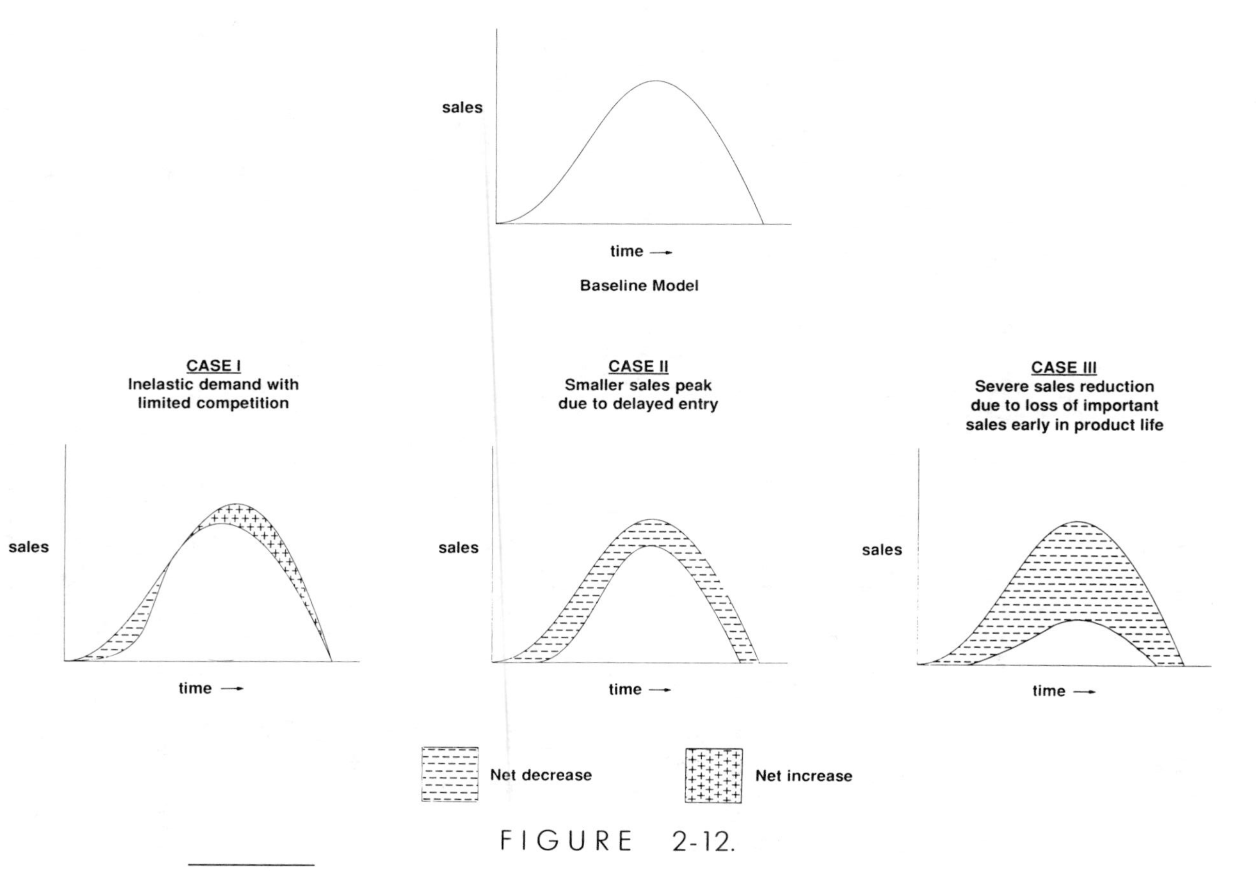

FIGURE 2-12.

The various effects on sales when the introduction of a product is delayed.

tions and average sales price assumptions. These are generally the most difficult to do well, and they tend to have a dramatic impact on the overall profit numbers.

Table 2-5 shows the profit impact calculations for a six-month delay. Once the baseline case and variations have been prepared we are ready to develop trade-off rules.

DEVELOPING TRADE-OFF RULES

Once the model is complete we can express each objective in comparison with the other ones. Ultimately, all of this analysis is worth nothing if we do not use it to change the way people behave. We have had success at changing behavior by giving people the tools to make decisions and helping them obtain the authority to do so. The tools we have found most useful are a set of decision rules that express changes in each objective in dollar terms. This allows comparing objectives with each other. For example, the analysis might suggest that accelerating the product's introduction by two months is equivalent to reducing its cost by 9 percent. In a case like this the design team might conclude that a one-month delay to redesign a subassembly, thereby eliminating 10 percent of the product cost, would be a cost-effective investment. On the other hand, a twelve-month delay would be excessive for such a small amount of cost savings. The purpose is to develop rough decision rules that let the development team make practical day-to-day decisions quickly. Should we build a second prototype for $50,000 if it would save us six weeks on the project schedule? If we know how much development time is worth we can trade off against this cost and answer this question.

The rules of thumb needed are developed by scaling our profit differences to easily handled values. For example, rather than saying that profits drop by 27 percent because of a six-month delay, we express this as a $2,814,592 profit decrease for a six-month delay, which equates to about $470,000 per month. If a 10 percent product cost overrun results in a $1,044,381 profit impact, this can be expressed as approximately $100,000 per percent of product cost overrun. Turning each development objective into a simple cost yardstick provides useful tools for making trade-offs (see Table 2-6). For example, consistent with the information above we could say that one month of delay is roughly equally damaging to a 5 percent product cost overrun.

TABLE 2-5

A Product Profit Model When Product Introduction Is Delayed by Six Months.

	Assumes:	Year −2	−1	0	1	2	3	4	5
Average sales price	−10%/year			$7,033	$6,330	$5,697	$5,127	$4,614	$4,153
Introduction price	$7,033								
Market size (units)				10,000	20,000	40,000	60,000	40,000	20,000
Market share			Lower	3.0%	9.0%	9.0%	9.0%	9.0%	9.0%
Unit sales			unit sales	300	1,800	3,600	5,400	3,600	1,800
Dollar sales				$2,109,900	$11,393,460	$20,508,228	$27,686,108	$16,611,665	$7,475,249
Unit Cost	−2%/year			$3,516	$3,446	$3,377	$3,309	$3,243	$3,178
Introduction cost	$3,516								
Cost of goods sold				$1,054,800	$6,202,224	$12,156,359	$17,869,848	$11,674,967	5,720,734
Gross margin dollars				$1,055,100	$5,191,236	$8,351,869	$9,816,260	$4,936,697	$1,754,515
Gross margin percent				50.0%	45.6%	40.7%	35.5%	29.7%	23.5%
Engineering		$2,000,000	$2,000,000	$1,000,000	$100,000	$100,000	$100,000	$100,000	$100,000
Marketing	16% sales			$337,584	$1,822,954	$3,281,316	$4,429,777	$2,657,866	$1,196,040
G&A	5% sales			$105,495	$569,673	$1,025,411	$1,384,305	$830,583	$373,762
Operating expense		$2,000,000	$2,000,000	$1,443,079	$2,492,627	$4,406,728	$5,914,083	$3,588,450	$1,669,802
Profit before tax (PBT)		($2,000,000)	($2,000,000)	($387,979)	$2,698,609	$3,945,141	$3,902,177	$1,348,248	$84,713
Cumulative PBT		($2,000,000)	($4,000,000)	($4,387,979)	($1,689,370)	$2,255,771	$6,157,949	$7,506,197	$7,590,910
Return on sales (PBT/ $ sales)				−18.4%	23.7%	19.2%	14.1%	8.1%	1.1%
Cumulative sales				$85,784,610					
Cumulative gross margin				$31,105,678					
Cumulative PBT				$7,590,910	Lower profit				
Average percent gross margin				36.3%					
Average percent return on sales				8.8%					

TABLE 2-6

A Technique for Developing Trade-Off Decision Rules.

Type of Shortfall	Cumulative Profit Impact	Period/ Amount	Thumb Rule
Six-month delay in product introduction	($2,814,592)	6 months	$470,000 per month
Fifty percent development expense overrun	($2,750,000)	50 percent	$55,000 per percent overrun
Ten percent unit sales reduction due to performance problems	($1,590,550)	10 percent	$160,000 per percent unit sales decrease
Ten percent overrun on product cost target	($1,044,381)	10 percent	$104,000 per percent product cost overrun

It is only when the model has been translated into these trade-off rules that it becomes a useful aid for decision making.

DOING IT

Now here is some practical advice on how to build a model in an organization. First, recognize that projected pricing is likely to be the most important variable. Second, pay lots of attention to the assumptions about unit sales with different introduction times. We have found that for certain products early introduction has a permanent impact on market share. Everyone recognizes the durability of the market position achieved by Lotus's spreadsheet program 1-2-3. In the medical devices market we have observed similar impacts such as that mentioned earlier in which introducing a product six months early meant the difference between having a 2 percent and a 20 percent market share. The design-in victories of OEMs often permanently lock out the competition, particularly when the first supplier has been clever about establishing standards that favor his or her product.

Use marketing heavily as a source of assumptions about these two key factors. Marketing people are normally the only source for such data, and although they can be quite reluctant to reduce their knowledge to numbers, once they have done so you will have obtained a key piece of the puzzle.

Once the input from engineering, manufacturing, and marketing is in, you have the raw material for the analysis. The beauty of this approach is that people rarely try to play games with the input until it is actually put into the model, by which point it is often too late to play games. The answers pointed out then by the model will usually be so compelling that minor changes in input values will have little impact.

In collecting input and building a model, different parts of a market may be found to behave quite differently. Build different models for them, and use the different outcomes to treat separate markets differently with regard to development objectives.

C H A P T E R 3

The Fuzzy Front End

In our consulting work we have repeatedly found rich opportunities to save time at the beginning of the development cycle. These opportunities deserve attention because, of all the actions described in this book, actions taken at the "fuzzy front end" give the greatest time savings for the least expense.

Yet these actions are frequently ignored because managers rarely focus on this stage of development. This lack of attention appears to arise from two sources. First, at the beginning of the process there are none of the traditional "handles" that managers use to control the process; there is no schedule, budget, or performance objective. Because there are no plans, activities cannot deviate from the plan, and deviation normally drives the management process. In a management culture that emphasizes management by exception, this stage of the process is not suited to be managed.

Second, this lack of handles is complicated by the fact that most managers perceive this stage as warranting only limited attention. As management has become increasingly financially driven, it tends to ignore this stage, which appears to have limited financial impact. However, this limited impact is only an illusion, as the financial analysis in Chapter 15 suggests. The traditional accountant will be misled by looking at the cost of this stage as being the cost solely of the people working on the project. Since this is rarely more than one or two part-time people, the cost appears to be quite small, often less than $20,000 per year. It is normally good management practice to focus more time on larger expenditures than small ones.

In reality, the true cost of this phase is usually many times higher than managers suspect. In this phase the most important influence on cost is the cost of delay, not of the manpower assigned to the project. The calculated cost of delay is often 500 to 5,000 times higher than the visible costs of assigned personnel. Managers unaware of these costs will tend to ignore the "fuzzy front end." Those who understand these costs will instead focus a great deal of attention on this phase.

This inattention could be excused if reductions of the development cycle could be achieved only with a great expenditure of financial resources. However, the situation is actually quite the reverse. The front end offers some of the cheapest opportunities to cut development time that are to be found anywhere in the cycle.

This chapter explains what the true sources of delay are in this early stage and what to do about them. As in much of the rest of this book, the approach here is different than is usual in other books on product development, which generally emphasize one of two issues. The first concerns selecting the right target, the right product idea to develop. The emphasis is usually on picking a product concept that is likely to win favor in the marketplace, independent of implications for the development process. A great deal of attention has been given to using techniques such as brainstorming and focus groups to identify successful products, with little thought devoted to completing these activities quickly.

Once a product idea has been selected, the second subject usually covered in the literature comes into play: how to develop a chosen idea into a product by using the imperfect organizations at our disposal. Now the issue of quick execution arises. It would be convenient just to confine ourselves to shortening the execution phase. The two issues are deeply intertwined, however, both because there are significant opportunities to shorten the idea-selection process and because the selection of the product idea determines the realities of the development process to follow. This chapter deals with shortening the front end, the next three chapters on shaping a product idea for rapid execution.

Planners often don't consider timeliness to be critical at the outset of a project, so time passes while a product's prospects for success are considered and reconsidered. Ironically, while the marketing analyses are being refined the market window begins to close and the chances of success continually decline. We will see that timeliness and product success are strongly correlated and can be achieved together. Nonetheless, achieving both requires conducting the product selection process in a way quite different from the normal one.

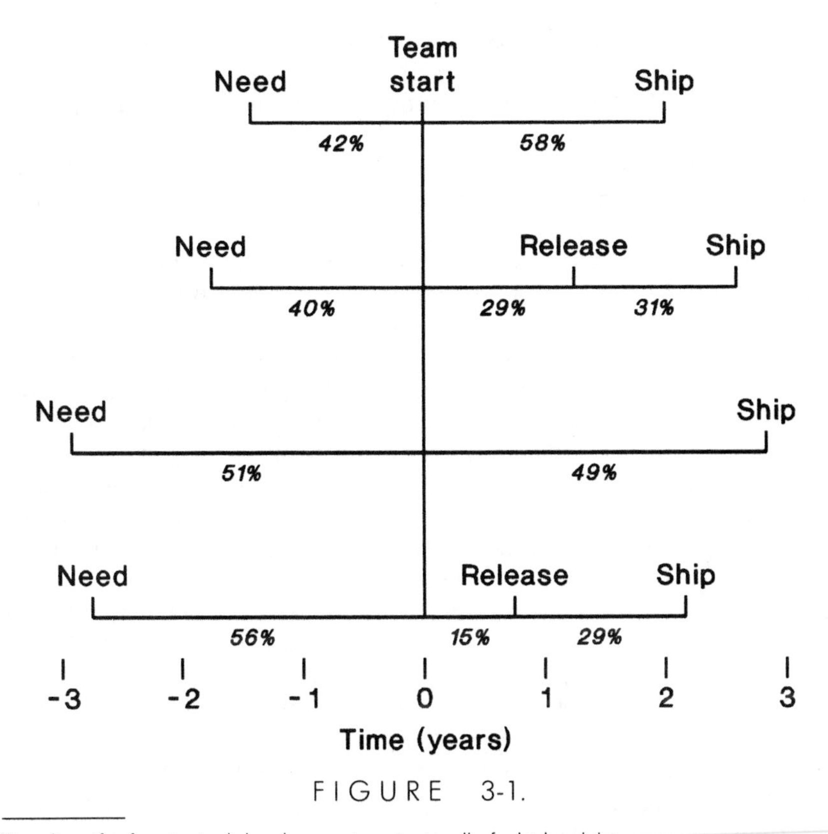

FIGURE 3-1.

Time lines for four typical development projects, all of which exhibit a long front-end period.

FUZZINESS CONSUMES TIME

Typical project time lines, such as those shown in Figure 3-1, can be surprising. Four events are listed on each time line: the time at which the product idea first became—or should have become—apparent to the organization, the time at which a full development team started working on it, the time at which engineering released the drawings to manufacturing (on two of the lines this event does not appear because manufacturing had continuous access to the drawings or else the release was spread over a long period), and the date when the first unit was shipped to a paying customer.

Notice the extraordinary opportunity to save time at the front end. Aligning the team start dates makes it possible to see that roughly half the development cycle has been consumed prior to assigning a team, even though the team start date is often remembered as the beginning of the project. This occurs because the front end is so fuzzy that people tend to forget that it even occurs.

The time lines illustrated are not in fact worst-case examples. In our consulting experience we have seen situations where as much as 90 percent of the development cycle elapsed before the team started work. In one case, a company sat on a new product idea for fifteen years, then initiated a crash two-year development effort.

Front-end time can slip away for many reasons. For example, a consumer electronics company that we know of has a line of high-tech products in which every new item generally features some type of electronic technology that is new to the company. The company's policy is not to develop basic technologies or electronic components within the company. In a typical development program the company identifies a new product and writes a preliminary product specification using a product planning process that takes six to twelve months. Then the biggest delay starts. The company does not have the resources to develop the new technology called for in the product plans, nor does it know where to find it. Because no one in the organization has specific responsibility to find the needed technology, the project goes into a dormant state until the technology appears. Projects can sit in this state for years until the key technology appears or a competitor's introduction spurs a more active stance in finding the technology.

This example suggests some of the countless ways in which front-end time can vanish. The frequently encountered factors illustrated in this example are that

- A key element of the product is new to the company.
- Clear responsibility is not assigned for this element.
- There is no system to prevent a project from slipping into a dormant state.
- There is no system to review dormant projects.

WHY TIME VANISHES

In this example it may seem quite apparent what the problem is and what should be done to overcome the delay. When one is in the midst of such processes, however, it is not so easy to see where the time is disappearing or what to do about it. Precisely because the front end of a project is fuzzy, it is difficult to spot delays. Later in a project the roadblocks become more visible as the development process becomes more concrete, there is a schedule to measure against, and responsibility for the project is clearer.

The Urgency Paradox

Because product development is driven by competitive pressure, much of the urgency to develop a product quickly comes from the marketplace or from fear of competitive product introductions. But the marketplace is fickle, and knowledge of competitive introductions usually comes too late. As a result, the signals to develop a product are poorly aligned with the market opportunity.

Each product has a certain time window in which the financial opportunity for it is greatest. Occasionally a product is introduced before this window opens, such as Studebaker cars, which appeared before the market could accept them. More often, the window of opportunity opens before a product is available. The market opportunity, although potentially large, often is not well defined at this point and the urgency for developing the product is usually low, because no competitive products are yet available. When working on a brand-new product idea people tend to assume that no one else is quite as far along on it as they are and that there is some breathing room. Consequently, when the product opportunity is the greatest, the sense of urgency is often the lowest.

Conversely, after a competitor has introduced a product the sense of urgency is high, but the peak market opportunity may already be passing. Because followers outnumber leaders, this is the more common situation. Much product development is done as a reaction to what a competitor has introduced.

So we have the urgency paradox illustrated in Figure 3-2: when a market opportunity is greatest, we are ignorant of it or complacent about it, but as the opportunity recedes our feeling of urgency rises.

The Market Clock versus the Project Clock

Product development is traditionally thought of in terms of cost. It is not only expensive to develop a new product but the cost of product development seems to be increasing faster than costs in general. This situation tends to make management cautious in committing to the development of a new product. Perhaps it is better to study the market opportunity more, they reason, before initiating full-scale development

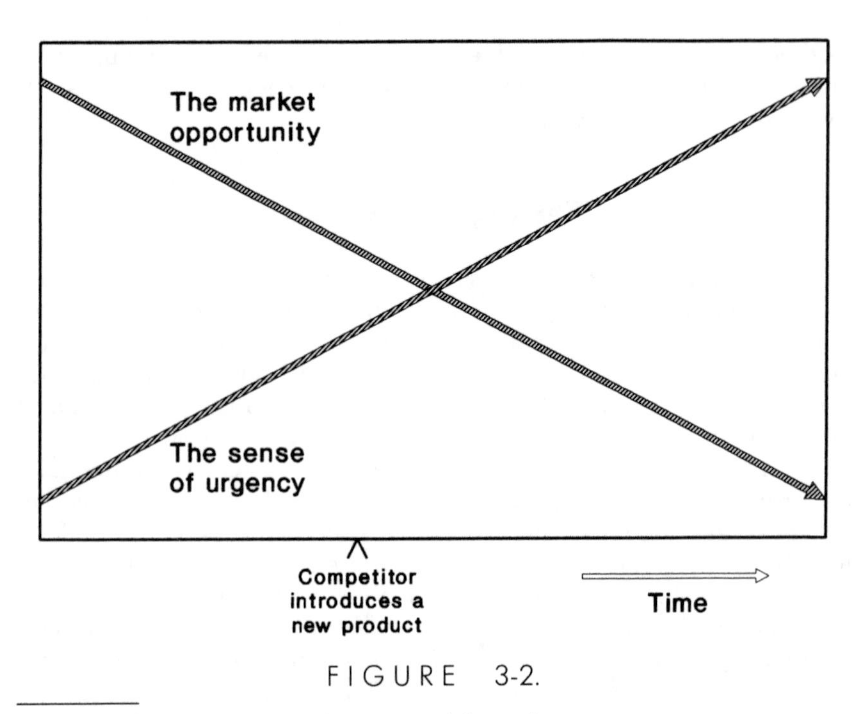

FIGURE 3-2.

The urgency paradox: a sense of urgency and the market opportunity behave oppositely.

activity. (This mindset is covered further in Chapter 12 under the concept of market risk.)

Management's reluctance to commit to a project is reinforced by the fact that spending rates escalate as a project proceeds into full-scale development. Figure 3-3 illustrates this effect. In the beginning of a project, only a few people are charging time to it, probably on a part-time basis; there are no big commitments or expensive supplies or equipment being consumed during planning. Toward the end, in contrast, many people in several departments will be charging their time to the project full time, components and machinery will be acquired, and management will be committing to advertising and tooling budgets.

Underlying this type of thinking is what we may call a project clock or "burn rate" mentality—which worries about activities in proportion to the rate at which they spend, or "burn," money. This attitude suggests that nothing is being lost on a product idea as long as little money is being spent on it. To this way of thinking the project clock starts to tick only when management assigns a team to it and stops ticking if the team is reassigned to another project.

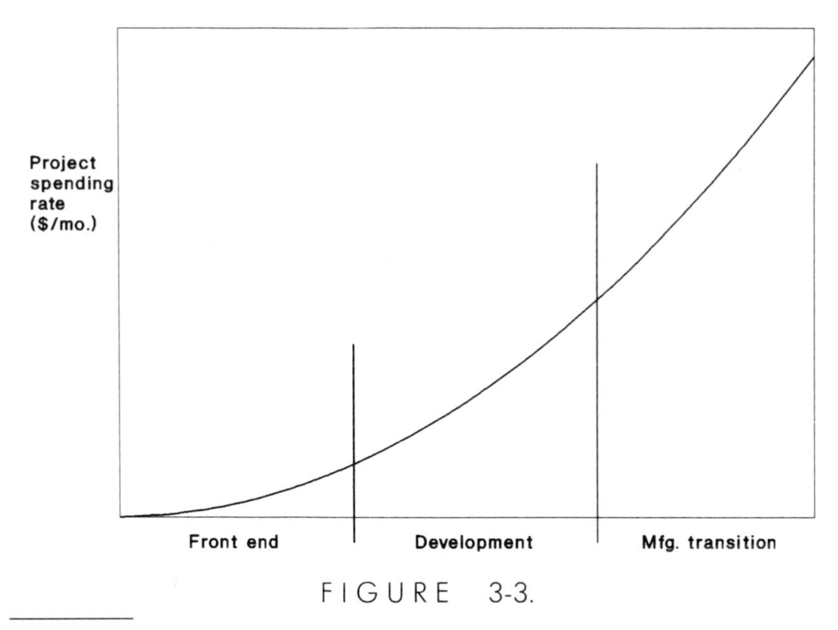

FIGURE 3-3.

A development project is inexpensive in the beginning, but the "burn rate" increases as it progresses.

We think of time differently. Rather than a source of cost, time should be considered an irreplaceable resource. This is the concept of the market clock, which starts ticking when the need for a product first becomes apparent in the marketplace, regardless of whether or not manufacturers then react to the need. Management has no control over the market clock; once it is started by an external event, it does not stop.

If management puts full emphasis on a project from the outset, the project clock's reading will agree with the market clock. But if the project is started late, is only partially funded, or is stopped, the project clock will lag behind the market clock. The amount of lag represents a lost market opportunity, as illustrated in Figure 3-4.

The market clock measures time that has monetary value. Chapter 2 described how to calculate the value of market clock time. The concept of the market clock is useful in many other ways, however, to keep us mindful of the value of time. For example, consider how priorities tend to be assigned to projects in general and product development projects in particular. The projects closest to fruition are the ones that usually tend to receive the highest priorities. Projects in their fuzzy initial stages get the lowest priorities, though, because they can't command the sense of immediacy that is clear for projects just days away

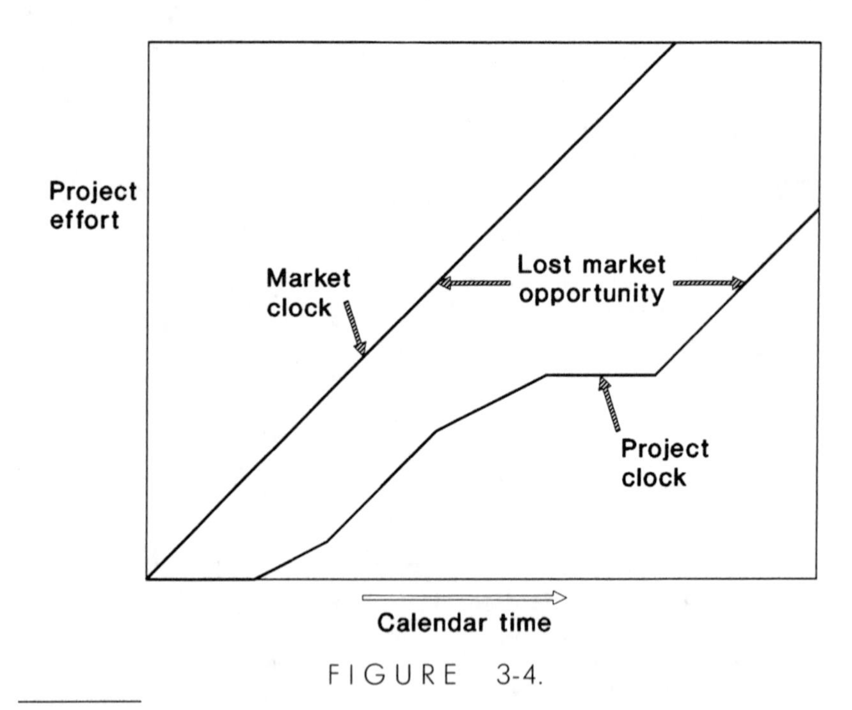

FIGURE 3-4.

Any lag of the project clock behind the market clock represents a lost market opportunity.

from being shipped. Our financially dominated management culture reinforces the tendency to emphasize the later stages of a project by focusing attention on projects with high burn rates. The market clock concept is helpful in reminding us that a week lost in a project is just as harmful in its first week as in its last.

HOW TIME VANISHES

Product Idea Initiation Processes

All companies have a formal or informal process, and sometimes many, for generating and screening new product ideas. For the most part these processes lumber along quite slowly, though occasionally they spring into action when a competitor introduces a product unexpectedly. Then a company can often react quickly with a flurry of product initiation activity. This vigor is often unrewarded in the marketplace, however, because the market opportunity has been passing while the reactive activity was starting.

One common way of initiating new products is the annual business-planning cycle. Unfortunately, annual planning processes do not treat new product ideas as the perishable property they represent. The first step in this ponderous process is usually to generate a list of new product ideas as input to the business-planning process. As the business plans move up through successive levels, product ideas are then screened progressively for strategic fit, financial payoff, and their operating and capital cost implications. Two types of delays plague this process. First, the business-planning cycle has objectives far broader than just evaluating a product idea. It attempts to integrate product plans with overall business plans. Although this objective is logical, it forces the product idea to sit idle while other parts of the plan are in process. The limited attention that a new product idea gets while it is in the annual planning process thus does very little to move the development process forward. The objectives of the planning process usually do not require that product ideas receive much evaluation in two key areas that affect product viability: their technical feasibility and market potential. These key evaluations usually have to wait until the project gets budget approval in the next fiscal year.

The second delay is the built-in one of annual planning. If a good idea appears perhaps in February of a given year, it may have to sit until, say, August, when the annual planning process starts. If the idea survives, it will then be approved for funding sometime during the following year. As a consequence, the idea will be delayed for up to a year by the need to fit it into synchrony with the annual planning process. Moreover, this process inherently puts all products on the same track. It is incapable of moving one program faster than another.

Another popular means of processing new product ideas is having a new products committee, which can be either a standing committee or an ad hoc one convened to consider a specific group of ideas. Standing committees enjoy the advantage of usually having high-level members. However, their weakness comes from the same source: these high-level members are too busy to meet very often or to give new product ideas the kind of detailed analytical attention they need to move along rapidly.

The ad hoc or task committee's strength is its ability to be customized to fit the circumstances. If it is formed properly it will have a specific objective, a clear deadline, and the type of support staffing necessary to consider the technical and marketing issues involved in assessing product feasibility. Compared with the standing committee, the ad hoc committee has the advantage of being able to impart a sense of urgency to a new product idea, to move decision making to a lower level, and of being more likely to involve those who ultimately will have to execute the idea, thereby building ownership in the concept. As discussed in later chapters, all these factors are important to accelerated product development because they help overcome the bureaucracy that by its very nature delays action.

The remaining way of handling new product ideas is the new products group. It can reside in marketing, R&D, or another area, such as with a vice-president of new products. Unlike the part-time new product roles inherent in other approaches, this group has the full-time responsibility of processing new product ideas. It should have the technical expertise needed, both in marketing and technology, to assess an idea adequately. The speed with which this group processes an idea is related directly to the job objectives it gets from above.

There are two problems with such a group. If the group lies in a particular functional area, such as engineering or marketing, it is likely to become polarized in its evaluation of an idea. Also, if it is given too much specialized functional attention at an early stage the idea is likely to have a poor technology–market balance or to succumb to being overstudied.

The second difficulty arises in either a functional or a cross-functional group and has to do with the very factor that is the strength of the new products group. Because the new products group is composed of specialists in the fuzzy front end, there is inevitably a handoff to the general practitioner when the decision is made to move the idea into development. These handoffs waste time, lose valuable unwritten product information, and undermine product ownership. In speedy product development handoffs should be avoided.

The Study and Planning Syndrome

Identifying and selecting a new product idea is only one of the activities that occur in the front end of a project. There is also analysis of its market acceptance and of the business opportunity it presents, planning for product line fit, and planning for financial and human resources.

Companies are now putting more effort into these up-front activities than they formerly did. A Booz-Allen & Hamilton study on new products found that predevelopment activities consumed only 10 percent of product development expenditures in 1968 but 21 percent by 1981. This increase is a reflection of growing awareness of the importance of new products, the cost of developing them, and the value of doing proper planning before execution. Thorough planning at the outset makes a great deal of sense. New product studies often conclude that the most successful products are ones in which the proper homework was done before starting development (see Suggested Reading at the end of this chapter).

The decisions that result from proper front-end planning will have great leverage. These early decisions can draw from a richer array of options, unconstrained by earlier choices. A decision made at the beginning of a project can apply to the whole project, not just the last portion. However, when a decision is made in the middle of a project it is likely to undo some completed work or at least weaken the project's momentum as people pause to familiarize themselves with a new set of goals. The proponents of heavy up-front planning observe that such early effort can actually save time overall because it reduces redirection and redesign later.

Even though there are good reasons for all the planning that occurs in the front end of a project, it nevertheless delays the project and thus adds market risk (see Chapter 12). The delay caused by planning, which is often a large portion of the development cycle, seems to be growing. As companies watch their budgets more closely in the drive to maximize quarterly earnings, they are more likely to "study" a development project than commit to it. When techniques of planning and analysis are applied improperly, they become a liability rather than an aid to management.

This does not suggest that a project should be started without proper study and planning. But studies and planning are indeed often misguided. These activities must be tempered by the recognition that only some things can be planned well at this stage. We advocate selective planning—concentrating on the issues that can be usefully resolved up front, and those that are valuable to resolve. When seeking additional information we should always ask, "Do we know enough now to begin on any aspect of this project?" If we do and don't act on it, we are creating unnecessary delay.

New product development is admittedly expensive and risky. But not developing a product—or delaying its development—can be even more costly. Some good methods for new product evaluation and analysis that have been developed in recent years should be used to keep risk to a minimum, provided that we recognize the risk of delay. It is not necessary to delay a project until all analyses of it are completed. The challenge is to get both the necessary analysis and the speed needed to be competitive.

HOW TO DEAL WITH FUZZINESS

A newborn idea for a product will always be fuzzy, but there are some things that can be done to keep the development process clear and focused on a deadline. Now we turn from understanding delays to minimizing them. Delays caused in the planning process receive attention first.

Moving Targets and Simplified Planning

The marketplace is always moving onward. If we have a two-year development cycle, we need to be working on plans for products to be introduced in two years, ones that will compete with whatever exists in the marketplace in two years. If, however, we can develop a product in just one year, the whole planning and forecasting process becomes much simpler, because our vision need reach only one year into the future.

Because the market is moving, fast product development and simple planning become mutually supportive. A quickly developed product needs only simple product plans, which allow the product to be developed quickly. Several phenomena interact in this loop. One is that long-range plans are not only more time consuming to create, because more factors come into play with them, but they are also less accurate. With less accuracy there is a greater chance that the planning will have to be adjusted later and the design be reworked accordingly. This process adds even more time to an already long loop.

Subdividing the Planning

We have just considered the planning phase as a single activity to be completed before product design can start. All too often it is assumed that all planning must be complete before execution begins, but there are many parts of the two activities that can be done concurrently. (Chapter 9 covers concurrency and overlapping activities in detail.)

The key to designing while planning is to realize that not all of the planning output is needed at the outset of the design phase. Fortunately, the planning data required to start the design work happen also to be the data that can be obtained the most quickly. This can be illustrated using a mechanical product with three stages of data and design refinement, as shown in Figure 3-5. This approach can be adapted to use with other products.

The first stage of data collection includes identifying the general parameters that position the product, such as the extent of its technical innovation, its price range relative to competing products, the intended user or customer, and its general size or weight. What is the vision of how this product should fit into the market? How does it relate to other

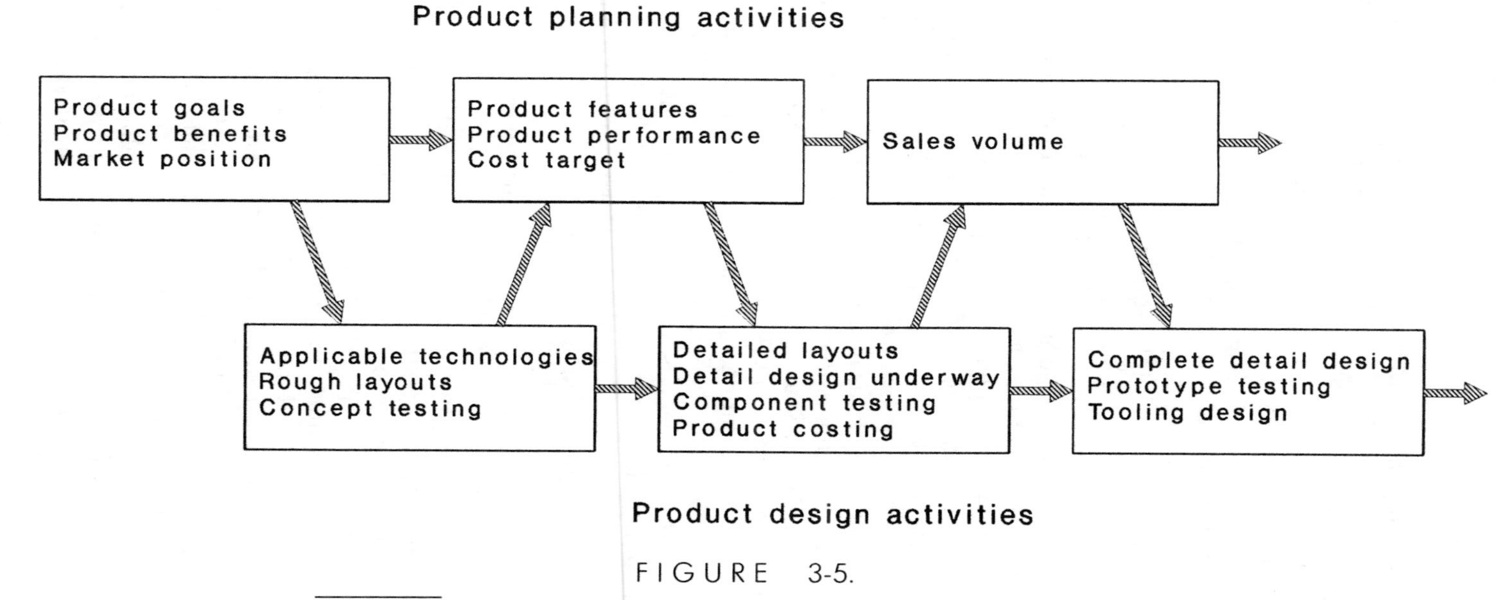

FIGURE 3-5.

Product planning and product design activities can occur concurrently rather than sequentially.

products the company produces or competitors offer? Why is it needed now? Must it have an image of being innovative or expensive, or is rugged reliability the key? The purpose of all this information is to get everyone who is involved at this point to thinking along the same general lines, to start the herd heading roughly east if that is the direction the company wants to go. The best way to focus everyone's attention in the proper direction at this stage is to choose a target customer and need.

The purpose of this information is to allow everyone—especially the designers—to start thinking about applicable concepts, available technologies, and potential areas of difficulty. Doing so simply narrows the solution space. At this point nobody should be putting specific dimensions or lists of features on paper. In particular, marketing should refrain from picking out the best features from each competitor's catalog to assemble a grandiose product (more about this in Chapter 4). It is more important now to establish an overall vision so that everyone can understand how the features have to fit together.

By now the design has already started in a sense. The next phase of information gathering is to collect detailed data on product features and performance and assemble them into a package to define the product. The common—and slow—way of doing this is for those involved to work in relative isolation, with marketing setting the features and performance by looking at competitive offerings or doing some market research. A faster approach is for marketing and engineering to make a whirlwind tour of key field sites to get this information directly from a few leading-edge users or customers. This technique is covered more completely in Chapter 5, on specifications.

Now the design people have most of what they need to design in detail. Most design and testing activities, particularly those on the critical path, can now proceed at full speed. Note that a precise sales forecast has not yet been generated. There is enough time in Stage Three to forecast sales for the new product. Generating a sales forecast for a new product is a slow, uncertain process subject to many unverifiable assumptions. If the product is unlike anything now available, there is little to compare it against, and if it is similar to the company's other products, one has to estimate the effect of cannibalization on these products, which creates a soft foundation in either case. Precisely because sales volumes are difficult to quantify, sales forecasts can consume a lot of time before everyone is comfortable with them.

It is important to recognize that accurate forecasts are not needed early in the development process. During development, forecasts have two uses. One is deciding if the economics of the product justify investing in its development, and the other is making design decisions concerning whether a particular part should be designed for low initial (tooling) cost or low unit cost. The first decision is usually the critical one.

Ideally, a decision to pursue the project, based on a sales forecast, should be made before any effort is devoted to the product. This decision can normally be made, however, with only a crude forecast. We have usually found that the best approach is to start with an educated guess about sales volume, then continually refine it as other information becomes available.

There is one further observation to make before we leave Figure 3-5. The information there flows continuously both ways between marketing and engineering. Each group provides the other with an ongoing stream of information needed for them to take the next step.

We have just considered a phased approach to product planning that can overlap with the design phase and liberate those who feel they must have everything planned out before they can begin designing. In particular, this method can help those who believe that all the market research must be completed before design work can start. Sometimes, in fact, the problem in an organization is just the opposite: there is no planning, and thus nothing really gets going. Let us examine this situation.

Building a War Plan

The military has long recognized that the time to plan for war is before war begins. They attempt to define likely military or political scenarios, then develop written plans to deal with them. If one of these scenarios should ever occur, they can implement the plan for it and have a large organization working in unison quickly. Though war plans are imperfect, they help to eliminate much of the confusion and consequent wasted motion that arise when an unexpected event occurs.

The same approach can be applied to preparing for an unexpected new product idea. When such an idea emerges, it often gets little attention because everyone is busy with their regular job. However, if a quick-react plan is put into effect, everyone's duties, responsibilities, and deadlines become much clearer and the new product idea can get a fast start.

The plan need not be an elaborate document. A task force of the key interested parties—perhaps marketing, engineering, manufacturing, and finance—should write the plan and have top management approve it. The plan should have an aggressive but doable timetable. On each day, specific activities of a marketing, technical, or financial nature should occur. At the end of the planned schedule a development project should be running, unless a decision was made along the way to abort it.

Just writing a plan and approving it accomplishes a lot. Writing it forces people to identify the key activities and put them in sequence. It implicitly says that other, unlisted, activities are not necessary to evaluate a new product idea. It alleviates the sense of indecision in which people wonder "what do we do now?"

Approval of the plan is where top management "signs up" for a new style of handling new product decisions. The war plan that has been created probably flows much differently from the normal process of dealing with new product ideas. Everyone is thus going to have to make some changes to follow the plan without delaying the process. By signing the plan, management indicates its willingness to be the first to change in order to enjoy the benefits of getting a product to market faster.

Writing and approving a war plan is valuable largely because it provides practice under benign conditions in cutting time out of a project's start-up phase. Just as military war plans are not envisioned to be executed as written, your plan is useful not so much for its specific steps as for its providing an acceptable framework to start a project quickly. In executing the plan, the key point—and the most difficult one, we have found—is for the general manager to choose about three people from key departments who can extricate themselves from their regular duties within a week and work full time on initiating a quick-react development program.

An actual example will show how a quick-react plan was constructed and demonstrate the pitfalls involved in executing it. This firm, a client of ours, serves the electronics industry. It normally takes them a couple of years to complete the fuzzy front end of a project, so they had formed a task force to write a quick-react plan. The general manager assigned representatives from marketing, engineering, and finance as well as an independent chairman, and charged the group with writing a plan that could be executed in sixty days. This was a clever group. After struggling with the work to be done, they decided that the boss must have meant sixty working days, not calendar days. Therefore on each of the sixty days they specified certain technical, marketing, and staffing issues that had to be resolved to initiate development on Day 61. In conformity with good practice, the plan concentrated on areas where this organization had experienced difficulty before. The plan was submitted to and approved by the general manager.

The plan was executed twice, each time with the same sad outcome. According to plan, the technical and marketing issues were in turn identified, addressed, and resolved. The staffing issues proved to be much tougher for this company, however, which is what we have often found to be the case. This firm was a division of a large, diversified corporation that did all product planning and budgeting in one annual process. There was no way to get the resources to start an unanticipated project (see Chapter 11 for more on this subject). Now the company is working on that issue. The moral is that when writing a quick-react plan, deal with the issues that are significant in your own company.

Keeping Fit

Competitive strength is built through fast product development by investing over the long term. Such investments include staying at the forefront in critical supporting product and process technologies and keeping abreast of market trends in the industry.

In many businesses, quick development can occur by adapting and repackaging known technologies in new ways to serve customers better. In other businesses these variations will quickly be exhausted and something truly new be needed. This usually means resorting to invention, which is a notoriously slow, uncertain process. Invention cannot be scheduled, so it is inappropriate for a tightly scheduled project.

The solution is to get the inventing done offline. Identify the basic technologies the business depends on and invest in creating a set of inventions in these areas before scheduling the inventions into products. For example, a company producing computer printers might be working on inventions in paper handling, printing, and servomechanisms as well as in manufacturing processes for making rollers or tracks more cheaply or accurately.

Similarly, develop a base of market information and spend some time studying market trends independently of any specific new products. In the process of executing the quick-react plan described above there won't be time to do in-depth surveys or collect much outside information. For the most part, you will be working with information already on hand when responding quickly to a new product idea.

In order for engineering to be developing key new technologies and marketing to be assimilating the proper information, both groups must understand the firm's master plan. The company needs to have a statement of what it plans to achieve with new products, which must be disseminated to the lowest levels of R&D and marketing. It may seem surprising even to make such an obvious statement, but we are often amazed to find companies that have vague long-term product goals, have no link between their product plans and an overall business strategy, or that have not communicated these plans to those who are supposed to be preparing themselves.

Staffing Considerations

Staffing is covered in detail in Chapter 7, but two points are pertinent here. One involves top management. Some CEOs like to be involved in the front end of a project where they can help shape fuzzy, abstract concepts, but many do not. They may be inexperienced in product development, or it may seem wasteful for highly paid executives to be working on an idea that may ultimately be discarded. Nevertheless, this is a company's best chance to couple its corporate mission directly with its product offerings, and is the CEO's strongest opportunity to make fundamental rather than rubber-stamped decisions on the product.

The CEO may help shape the product, but as a product concept emerges from the fuzzy front end it will also need a full-time attendant to provide continuity with the future development team. Much time and information will be lost if there is a total handoff when the development team is assigned. At least one person should follow the product from the front end into development.

SUGGESTED READING

Cooper, Robert G. 1986. *Winning at New Products*. Reading, Mass.: Addison-Wesley. Much of the marketing-oriented new products literature concentrates on evaluations needed at the front end of the development cycle to enhance the prospects for product success. This book is a good example of such literature. The trick is, of course, to conduct the evaluations with the same sense of urgency that exists at the tail end of development or to do them concurrently with initial design activities.

CHAPTER 4

Innovating Incrementally

To get new products to market quickly there are two issues to be addressed. First, we need to minimize the amount of work to be done to develop a product, then find a way to get that work done effectively. This chapter covers the issue of controlling the size of the product development job.

Product development involves much that is new and thus requires a lot of learning. One has to learn what product features are needed, which design solutions will work and which won't, what can be provided by suppliers, how various manufacturing capabilities can be employed, and countless other details. Some of this information can be learned through straightforward analytical processes but some learning, in true inventive style, requires that mistakes first be made. Product development time is basically learning time.

Unless the amount of learning that must be undertaken is managed, it can be overwhelming and make the development project take much longer than desired. Much of the learning is routine and necessary and can be handled without delay. Some, however, can consume a great deal of time because it concerns new items that must be understood or factors that interact in complex new ways with the familiar ones. The astute manager watches for learning situations that will cause undue delay.

Part of the art of fast product development is understanding the magnitude of the learning burden associated with a given project and knowing when to draw the line on one project and defer the remaining learning to the next one.

THE MEGAPROJECT TRAP

Product development people, particularly in North America, like to be involved in big, important development projects. The megaproject gets more attention and money and presumably has a greater economic benefit. Mounting evidence argues that this emphasis is an illusion. Nonetheless, it seems to be a common problem. Other factors being equal, Americans will opt for an all-new blockbuster product. In other countries, specifically Japan, a series of small wins seems to be more popular.

The difficulty with the big project is that it carries an enormous learning burden, which means there will be a long and unpredictable development period. Project difficulty is deceptive, because it does not grow in proportion to the number of new items. Instead, it grows by compounding. Each new element usually has to interact with a number of existing ones, so as the base project grows bigger the addition of a new element becomes increasingly difficult. For example, going from three product features to four nearly triples (from four to eleven) the number of potential interactions between features, each of which might derail the project.

Often it is more efficient to break a big project into two smaller ones that can be completed faster. Extending this line of thinking, the most effective way to make progress quickly can be to take many small steps through incremental products. As noted, this approach seems to go against the grain in North America, but the Japanese have used it to great benefit.

Stalk and Hout, in the book referenced in the Suggested Reading for Chapter 1, provide an excellent example of innovating through small steps to gain a large, cumulative advantage. Mitsubishi Electric upgraded its three-horsepower heat pump dramatically over a five-year period by making incremental but coordinated changes annually. In the first year, integrated circuits were added. They were then replaced by a microprocessor in the second year, with a couple of significant installation convenience features added. In year three, Mitsubishi replaced the reciprocating compressor with a more efficient rotary one and improved condenser efficiency. Changes in the fourth year concentrated on enhancing the control system by adding sensors and exploiting the power of the microprocessor added two years earlier. The new compressor's motor was the focus of the fifth year's activities, in which an inverter was added and tied in with the previous year's control system to automatically control motor speed and further raise the unit's efficiency. In five years the whole unit had been redesigned, but with changes introduced annually rather than in one "revolutionary new product" (with permission from The Free Press, 1990).

The "Do Everything" Product

Having a revolutionary new product is an alluring goal. For a company that has fallen behind its competitors in product innovation, a revolutionary product is even harder to resist. An all-new product may seem to be a necessity because, if only a few minor changes are made, their company will still lag behind the industry leader, and the competitors' progress has demonstrated that big improvements are feasible.

The classic do-everything product comes about when someone, typically in marketing, studies several competitors' catalogs and starts making comparisons. The temptation is always to pick the best of each and combine: use the product performance advantages from competitor A, combine them with the attractive styling of competitor B and the user-friendliness advantages of competitor C, then dream of selling it at the price of competitor D.

Even though do-everything products are routinely planned this way, this approach will not stand up well in the cold light of analysis. If we were to disassemble product D we would likely find it to be inexpensive because its creators had purposely made some compromises. Perhaps they decided on a utilitarian housing because they needed to keep the cost of the plastic resin in it to a minimum, or they had to use a rather primitive housing shape to minimize the complexity and therefore the cost of the molds for it. Cost may also have limited their electronics budget, affecting their ability to provide much in the way of user friendliness or advanced performance.

The do-everything product may be unrealistic, but this will not diminish people's desire for it. It tempts each function in a different way. Engineering is tempted by the technical challenge of doing something that has never been done before. The engineer can experiment with the newest technology to see how, or if, it applies. If a real breakthrough can be accomplished, it will look good on one's resume and be more likely to lead to patents, further satisfying the ego. Time to market takes a back seat when one starts thinking along these lines.

Marketing's desire for a do-everything product revolves around the advertising claims that might be made for it. Think of the many all-new features to be listed on a catalog page. An all-new product is more exciting to promote than one that is just slightly better than last year's model, especially if the slightly better product still falls short of something a competitor is offering.

Manufacturing prefers that product changes be lumped together to minimize production interruptions. A new product disrupts manufacturing's routine output, which can influence their pay, based on normal production measurement systems. Each time a new product is brought on-line, production has to be stopped to make pilot runs, new parts must be added to inventory, perhaps with new suppliers, and learning curves start anew as people are retrained for new tasks. The compensation system usually does not reward manufacturing people for these headaches, so the less frequently they occur the better for them.

The do-everything product is what comes naturally; incremental innovation would not win a popularity contest. The only way to shift toward a small-wins mentality is to reeducate participants and continually reinforce two messages: that rapid product advancement is important to the company, and that incremental improvement is the fastest way to improve products.

The do-everything product is a trap that lures everyone. Developing an all-encompassing product is a big, unpredictable effort, which pushes companies toward big development projects. Big projects are themselves a trap that slows innovation.

Giant Steps Demand Even Larger Giant Steps

Product development is risky, and the risk mounts as the development task grows. Management often copes with increasing risk by undertaking even larger tasks, which increases risk even further. Consider the chain of relationships that lead into this ever-worsening spiral.

Suppose that you want to develop a product to replace one in your current line, which is a common starting point for new products. If this product has not been upgraded for several years, you will need to make some major improvements. It will probably be several more years before you will have the opportunity to introduce another upgrade. Consequently—and this is the key point—you have to get this upgrade exactly right because you will have to live with it for several more years. There is a big job to do because of past years of neglect, and because you appreciate how the system will work in the future you have to do the job just right. We are now starting around the loop illustrated in Figure 4-1.

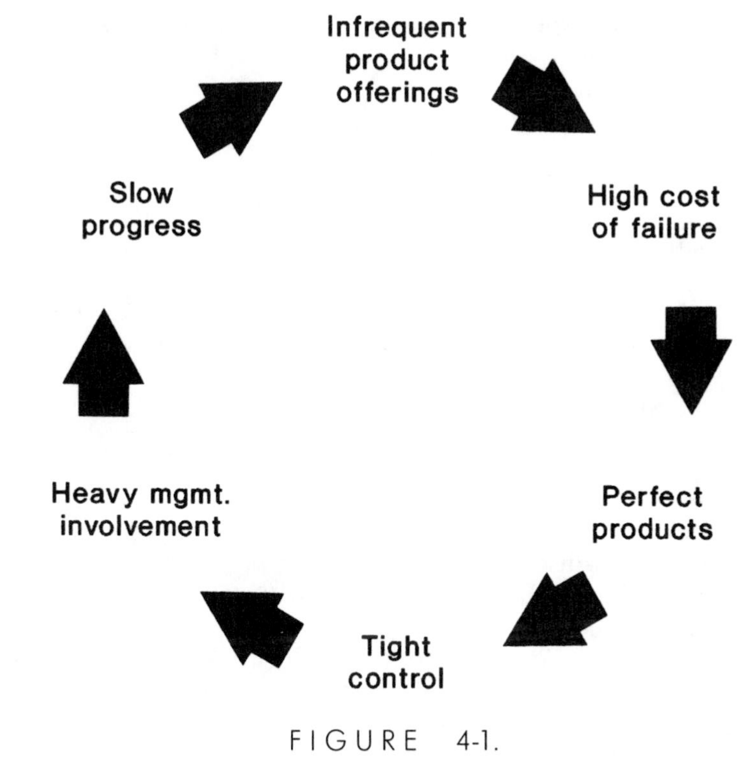

FIGURE 4-1.

Long development cycles encourage even longer development cycles.

Product development has many pitfalls, but because this project has become important to the company, it can't be allowed to fail. So you take two steps. You put several controls in place to manage progress. These controls include periodic written progress reports, project reviews, and engineering drawing control procedures. You also make sure that senior management is regularly involved in project decision making to discover problems and offer advice out of their years of experience.

Tight controls and heavy management involvement achieve their objectives, but at a high price. Both measures slow progress and delay introduction of the product. You are lucky if the market, over which none of your controls has any influence, has not shifted by the time the product is finally introduced.

By now you will have invested so much in the product's development that you won't be able to touch it again for several years. But other products are now desperately in need of upgrading. And thus the cycle continues.

The Costs of Complexity

The message is that as more and more requirements are added to a development project, it gets more complicated and the project takes longer. To really make the point, however, let us see just how much longer it is, quantitatively. As suggested earlier, project complexity grows by compounding, rather than simply growing linearly as more requirements are added. Putting in a few more product features or cost restrictions can add a lot to the schedule.

Project complexity compounds in several ways. First, as new elements are added to a large project, they have to interact with more existing elements than they would for a smaller project. Each new element creates more interactions and thus more work as a project grows larger. Figure 4-2 illustrates how the number of potential interactions compounds as the number of elements increases.

An element here is simply any convenient measure of the basic units of the product. For a mechanical product the number of parts or fasteners could be counted. In electronics it could be the number of components, circuit connections, or boards. For software one might count lines of code or memory units. Or one can work on a higher level by counting subassemblies, modules, or features as elements.

Figure 4-2 indicates only the potential for interactions, for seldom do more than a fraction of the elements in a product actually interact. Nonetheless, the potential indicated for mushrooming complexity is clearly evident. Each new element roughly doubles the potential number of interactions. It adds as many potential interactions itself as existed among all the elements before it was added. For example, nine elements offer 502 potential interactions, and the tenth adds 511 more, for a total of 1,013.

The second way in which a project can get more difficult with increasing size is when each element must be designed more carefully in a large project, which means more work per element. A large project presents a more complex environment in which more can go wrong, so each element may have to be designed under tighter constraints to maintain a given level of system performance.

This general concept works out differently in each product, but viewing it in terms of system reliability provides a simple way of quantifying it. Suppose that we need a system to work 99 percent of the time and it works only if all its elements work (this model strictly applies only in certain situations; for example, you need an engine and

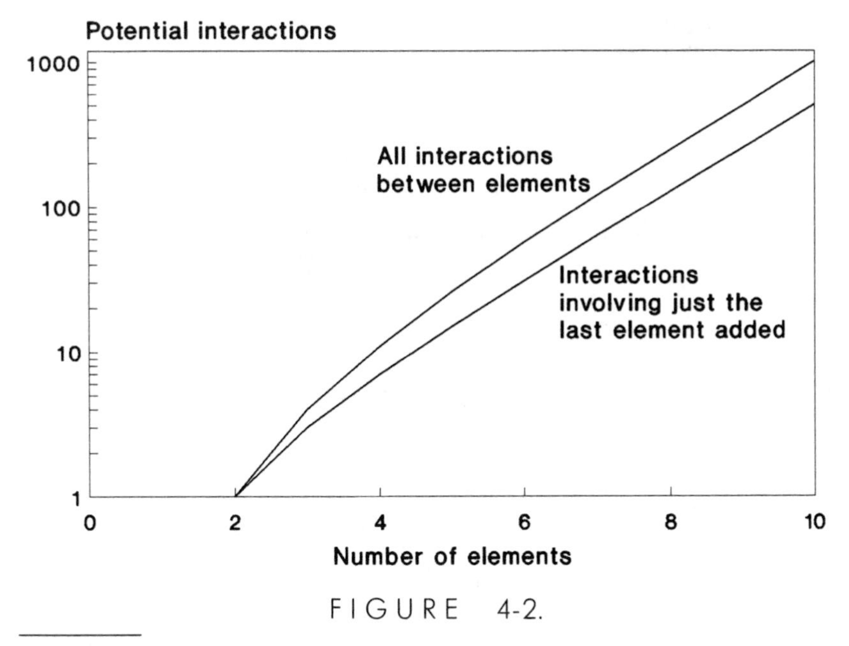

FIGURE 4-2.

The potential number of interactions among elements approximately doubles as each new element is added, and the potential for complexity rises accordingly. (The equation for the upper curve is

$$N = \sum_{i=2}^{n} n!/[(n-i)!i!]$$

where N is the number of potential interactions and n is the number of elements.)

brakes but not necessarily a radio or bumpers to use a car). Figure 4-3 shows how reliable each element must be to provide a constant level of system reliability as a product grows more complex. As the system grows, each element must be designed to higher standards to maintain system goals, requiring more work, time, and probably cost per element.

The time penalty for adding a new feature is very real, whether or not anyone recognizes it to be. To understand why, think about how Figures 4-2 and 4-3 apply to daily operating issues. Often a manager or a marketing person will ask a technical expert such as a project engineer if a certain feature can be added to a product without much disruption. Sometimes the technical person says no because the addition will require changes in basic components or a fundamental change in system architecture, which is a quantum jump. But often a technical

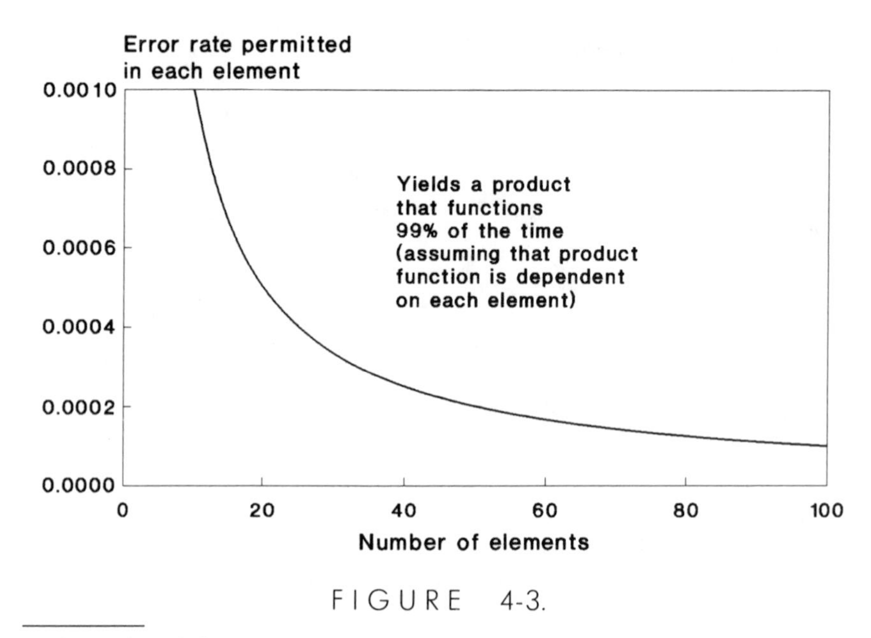

FIGURE 4-3.

As the number of elements in a product grows, each element has to be designed more and more carefully to maintain overall system goals. (The equation used is $E = 1 - 0.99^{1/n}$ where E is the error rate per element and n is the number of elements.)

expert will be accommodating and say that the addition can indeed be made without causing a significant disruption. Figures 4-2 and 4-3 thus apply together in that an addition can in fact be made within the current design framework, but it will drive the project to the right on both diagrams. The addition will require checking more interactions among elements and mean being more careful about the design of each element, including the existing ones, to preserve system performance.

We conclude that adding a few elements can complicate a design to a much greater degree than is usually appreciated, which can ultimately have a great impact on the schedule. To make this point graphically, we picked one particular situation to use as an example in Figures 4-2 and 4-3. These figures relate to specific system models that may not represent another product well. Their numbers should not be used to design a system, but they can be useful for clarifying the concept of product complexity to make it possible to build some applicable models or at least know the proper questions to ask. Through astute system design (see Chapter 6) it is often possible to avoid, for example, having a system's operation depend on the operation of each element. But these alternative designs may themselves add time to the schedule or add cost as, for instance, by using redundant elements to overcome reliability problems.

Furthermore, the two factors illustrated in Figures 4-2 and 4-3 are not the whole story because there are other ways in which product complexity compounds. For example, the sheer number of untried innovations adds to complexity. These innovations can include alternative suppliers, new manufacturing methods, new operating conditions and environments, and new materials. Although being able to innovate depends on being open to these options, the effective innovator works to narrow the solution space as quickly as possible and at an appropriate point draws the line on exploring further options.

TECHNIQUES OF INCREMENTAL INNOVATION

The concept of incremental innovation calls for introducing small but frequent changes in a product under the guidance of a product master plan and using customer feedback to guide future steps. Several techniques are useful in shifting into this perhaps unfamiliar mode of operation.

Take Quick, Frequent Steps

The essence of incremental innovation is to complete the development cycle as many times as possible over a given period. The development process is like a game in which the players learn a little more about how the game works each time they play it. Each time an organization completes a development cycle it receives a bonus payment of learning, which can be spent to make a stronger start on the next cycle. In the end, those who complete the most games will have the best skills and know the most about the marketplace.

In automobile development the Japanese have a considerable cycle-time advantage over North America and Europe. Kim Clark and others at the Harvard Business School have studied product development in depth among the world's auto makers. They have found that on average it takes Japanese companies forty-three months to develop a car compared to sixty-two months in North America or Europe, for a ratio of about three cycles in Japan to two elsewhere. Projected out, this means that the Japanese can introduce and gain experience from nine generations of products while we are left with only six.

Contrary to the faster–smaller steps approach, the Japanese steps in automobile development happen to be faster but not smaller than U.S. steps. Therefore, the Japanese gain a clear-cut victory every time they get another generation ahead. But even if their steps were smaller, they would still have an advantage because of the learning that takes place from introducing a product. This process allows them to aim future product generations more precisely.

The original Sony Walkman, introduced in 1979 at $200, was not at first a very attractive buy, but Sony soon made it so by introducing dozens of small changes quickly and sequentially. These steps included manufacturing improvements aimed mostly at cutting product cost and improving its performance. Many of these changes involved replacing mechanical components with lighter, more rugged, cheaper electronic items. As the product's price dropped, expanding its number of potential buyers, countless varieties were introduced to test the marketplace. This was learning by doing, not by planning and studying.

Experimentation—in the marketplace—needs to become everyone's operating style. Japanese manufacturers tend to view each product as an ongoing experiment that they are constantly trying to improve. There is no perfect product and never will be—the only question is how closely a firm can track a moving target in the market.

Most manufacturers, even slow ones, do a lot of experimenting, which is a key to innovation. The critical difference is that most manufacturers don't do their experimenting in the marketplace. Instead they do it through studies, plans and design options conceived and refined thoroughly in the development lab. To move quickly, these experiments have to get out of the lab even if they aren't yet perfect. Tom West, the team leader in Tracy Kidder's *The Soul of a New Machine*, had one prerecorded message he uttered as part of his misson: "Okay. It's right. Ship it."

Just how a company chooses to make quick, small steps is an individual matter to be worked out as part of a company's strategy and values amid the demands of its markets. There are no standard prescriptions, but here are some examples.

Chrysler initiated the minivan concept in 1984 by introducing its Caravan and Voyager models with the so-called S-body. This vehicle had been conceived in 1978, when gasoline supplies were a concern and prices were high. The design goal was thus to produce a light seven-person van with a fuel-efficient 2.2-liter engine. However, by the time the S-body was to be produced, gasoline prices had abated and a power–performance theme was being heard in the automotive market. In a sense, Chrysler had the wrong design for the times. There was certainly a case to be made for scrapping or at least heavily modifying their economy model design before committing to high tooling, plant start-up, and market-launch expenses. Chrysler's response was, however, to proceed with their design but incorporate a larger engine later.

Ford was reaching a similar decision point at about the same time with regard to its Taurus/Sable project. This model too had been conceived in the late 1970s as a small economy car, but when the market shifted in the early 1980s Ford decided to scrap the original design and start over on a bigger car. The Ford products finally appeared in 1986 with four-cylinder and V6 engine options.

Meanwhile, Chrysler had captured the minivan market in 1984 with its original small, four-cylinder design. It then followed up, in 1987, with a longer-bodied V6 version. With both vehicles it now holds a commanding two-to-one sales lead in the popular minivan market over comparable GM or Ford products. Chrysler established a market with an imperfectly positioned product, then maintained its leadership with a responsive follow-up. In other words, they took two small steps instead of one "perfect" one.

The next example comes from an electronics firm client. Although this company has difficulties developing the electronic innards of its products, we found a big delay in developing the product's housing and control panel. Styling and ease of operation are important sales features, so a great deal of effort goes into the relatively subjective areas of ergonomics, colors, textures, and graphics design, all decisions that are influenced by everyone from the chairman of the board to the sales force. Housing design turned out to be the gating activity in their development schedule.

The solution in this case was having two small steps in housing design instead of one big one. The company starts by designing a first-generation housing and control panel, which they freeze while they concentrate on debugging the electronics. Then a refined second-generation housing is developed as a running change after production starts. This is a custom solution employing incremental innovation that gets an acceptable product to market quickly, then follows it with a handsome one a short time later. More important to the long run is that this two-step procedure introduces their people to the idea that the first product introduced need not be the ultimate solution, which prepares them to invent more powerful incremental innovation solutions in the future.

One last example relates to the 1991 model Buick Park Avenue, which was being planned by the end of 1986. This car was to be a major facelift, including a new body, significant engine and transmission enhancements and a whole group of suspension and steering improvements. The program planners also wanted to minimize design complexity, because they needed a short development cycle to meet a March 1990 target for starting production. Their solution is interesting because it is so counterintuitive. The planners pulled ahead many of the suspension and steering improvements for introduction during the 1989 and 1990 model years as their Dynaride package. With a few adjustments to struts and engine mounts, Dynaride was then available as an existing element in the 1991 model.

Learn from Customers

If we knew just what kind of products customers wanted and how much they would buy, incremental innovation would have limited value. It would then be used only to cut corners because of suspicions that a competitor was at our heels. An unrefined product might then be better than a late one.

But we don't know what customers want, and often they don't either. It is too easy to fall into the trap of designing them something they won't buy. Here is where incremental innovation can save lots of time. Design something that is minimally acceptable, check it out by selling it for a while, then adjust and readjust the product until customers say it is right.

In this age of sophistication and high-powered computer analysis, it is easy to delude ourselves into thinking we can lay a consumer's psyche out on a spreadsheet. Market surveys, focus group data, and conjoint analyses are all available by the ream. The fact is that we know very little about what a customer will actually buy, especially if we are developing something significantly different from what already exists. The state of the art is well captured in Lesch's Law, attributed to the former CEO of Colgate-Palmolive: "You never know what is going to sell until you try to sell it."

Companies spend a lot of development time trying to second guess the customer when they could find out what they need to know firsthand. For instance, Black & Decker is a leader in eliminating cords from electric tools and appliances by replacing cords with rechargeable battery packs. Their Handy Mixer was among the first items in a line of kitchen appliances designed to capitalize on cordless operation. Their presumption in designing the Handy Mixer, even after introducing several cordless tools and vacuum cleaners, was that users expected the cordless product to be as powerful as the corded one. Their primary emphasis was thus on developing a cordless mixer with performance nearly equal to a corded one, even though it became larger, heavier, and considerably more expensive than a corded mixer. After considerable discussion, Black & Decker decided to develop an "underpowered" version as well because a few people believed it might also have a small market. In the end, this minority was vindicated by the marketplace. Reports from people who bought the units indicated that convenience and light weight were its strong points, and they didn't expect the small model to be a heavy-duty mixer. The larger, more powerful unit meanwhile met with a poor reception in the market and was soon discontinued, because people saw little point in paying a premium for performance only equal to the corded unit they already had.

Because we lack the information really needed to develop products, we often make presumptions about how customers will react. It happens in every industry, as this example from RHEOX (formerly NL Chemicals) suggests. RHEOX's chemists had developed a rheological additive for unsaturated polyesters. This material thickens liquid resins so that they will retain position when sprayed into molds to manufacture items like shower stalls. RHEOX's additive had cost-effectiveness advantages over competing products on the market, but had one drawback: it required an extra on-site mixing step before it could be stirred into the resin. Marketing perceived this extra step as being unsatisfactory to customers. Then R&D discovered an additive that did not require the extra mixing step, but did require RHEOX to invest in a new factory machine. This created a standoff because the capital investment could not be justified until enough customer demand was indicated, and without the new machine the demand could not be demonstrated.

The solution was to conduct a field trial with the extra-step product that they could make. RHEOX's vice-president of research, Michael DeSesa, displays here his bias for action: "Always push for a field trial. It's a no-lose situation for marketing. If customers don't want it, marketing can say, 'See, we told you the product didn't have all the attributes we said it had to have.' On the other hand, there's nothing like a customer saying, 'Hey, this stuff really works like you said it would, and I want to buy some.' Now the roadblocks begin to crumble." Armed with successful field-test results, DeSesa was quickly able to get authorization for the new machine. By adapting their plans to get intermediate feedback from a customer who had used the proposed product, RHEOX was able to accelerate their innovation process.

These examples illustrate just a few cases among many that could be given of situations where a product's evolution has been advanced by getting the customer involved. Companies like Toyota and Panasonic (Matsushita) go much further than this in developing customer feedback into a competitive tool. For them the real power of customer involvement comes from coupling a prolific development process with ongoing customer sampling. Rather than relying on market research's mixed record in pinpointing customer needs, they instead continually offer a variety of novel products and watch what customers buy or don't buy. The faster their development process can react to how their products are selling, the stronger they become as a competitive force.

One word of caution is in order here. Listen to your customers on how often and where they will accept change. With consumer products this isn't usually a serious problem. A customer may be surprised upon revisiting a store to find that the video recorder he or she purchased last month has since been superseded by another, but this will not affect their future actions. By contrast, industrial customers often have to pay for changes in their own systems to incorporate a new product, so keep tuned in to their tolerance for change.

Make Technology Transitions Transparently

Because of the urge to undertake megaprojects, whenever a new technology with great potential comes along, product developers push to exploit it fully in the very first product. But this is a slow and hazardous route. It is not easy to transform a product development group's skills and thinking patterns from, say, mechanical control of a machine to microprocessor control, from electrical transmission of signals to optical, or from sheet-metal fabrication to molded plastics.

There are two stages of learning needed in any such transition. One is to learn the basics of the new technology so that it can be made to work in an intended application, under actual production conditions. The other is to learn the finer points, the potential trade-offs, and the techniques for increasing the leverage of the new technology.

In incremental innovation we need to make a clear distinction between these two steps. Shifting to a new technology is a big step in itself, bigger than a company should normally be taking. Even when it is implemented as simply as possible, moving to the new technology may prove to be a more complex project than should be attempted by a fast development team.

Once the new technology has been implemented in a first, rather rudimentary, product, the company is in a strong position to start extending it. Production learning experience will then cut costs and improve yields as people learn how to work with the new technology. Product developers will be able to get information back from customers and product service centers to guide future extensions. Focus groups will become more meaningful with a realistic sample of the product to show participants. In general, the company won't waste its efforts pushing the technology in directions that are of no concern to consumers.

It takes a great deal of marketing and technical restraint to introduce a new technology in a minimal form. Many extensions will appear easy to achieve, and the new product may be hard to sell at acceptable margins unless it offers at least some potential enhancements immediately. To combat this natural tendency, it is useful to think of making a transition to the new technology that is transparent to the user. Sneak into the new technology with a product that is as similar as possible to an existing one. Don't promote the new features; instead, accept reduced margins if necessary on this transitional product.

Observing restraint on the first new-tech model and introducing it relatively early compared to those of competitors leaves a company poised for expansion into the new area. This is the time to get heavily involved with customers so as to work their desires into follow-up variants. If an entry model, which purposely had limited benefits, proves not to be a business success, then after there are a couple of alternative models either gracefully drop the entry model or redesign it to improve its cost–benefit characteristics before it ends up costing the firm too much.

Knowing when to move into a new technology is perhaps the most difficult strategic decision a company has to make. Genuinely new technology generally is not suited for the fast-react teams we suggest for other, more routine, product advancements. Yet new technology has to be employed, in order to remain competitive. Our solution is to get into the new technology as quickly and simply as possible, and then—but only then—shift to incremental innovation to enhance product features and margins.

IBM provides some excellent examples of making technology transitions transparently. When a new semiconductor chip becomes available to IBM, there must surely be a strong urge in this company renowned for its technical and marketing clout to put the new technology into a "revolutionary" new product. But note how they introduced a few recent technology breakthroughs. The first example is the initial application in the computer industry of the Intel 80486 microprocessor chip. This could have been billed as a big deal, but IBM merely used it to substitute for an 80386 chip to upgrade the performance of an existing line of personal computers. Similarly, IBM announced, on July 26, 1989, the first application in the world of a four-megabit memory chip. They could have built a sophisticated new computer around this chip, but that would have taken too long to get into the market and traverse the learning curve. So IBM merely stuck the new technology onto a lackluster memory expansion card that plugged into their top-end personal computers.

It almost seems that IBM built its Personal System/2 line of personal computers with a flexible architecture thinking of using the system as a testbed that could be used easily to get started launching new technologies. This nearly transparent transition may not have been a glamorous or cost-effective way to use the most advanced technology available, but it did place IBM ahead of the pack and position them to gain value from the new technology.

Freeze the Concept Early

As with much of what we suggest throughout this book, incremental innovation involves making an agreement, a contract. This understanding recognizes that two processes must work together in order to be able to innovate quickly. One process is product development. In order for the development process to move ahead rapidly, it must have a fixed target and not change course. The other, interconnected, process is market forecasting. In order for a forecast to be reliable, it must work with short horizons. If a new product is not delivered quickly the market shifts, requiring forecasts to be revised and design goals to be changed, and the development process stretches out, thus opening the possibility of yet another shift in the market.

The way to get out of this circular trap is to agree that at a certain point the design is to be frozen. With this simplification in their job designers then agree, as their part of the contract, to get the product out on schedule. There is still a chance that the market may shift, but marketing must agree, as its part of the deal, to accept this risk in exchange for a short development cycle that will greatly simplify their planning process.

Each company needs to pick a freeze point that works best for it. There may still be times when the design will have to be changed after being frozen, but these occasions should be clearly limited to issues such as product safety or reliability. If the color is not quite right, the controls not as easy to use as expected, or a new feature is identified after the freeze date, it must be held for the next product, or be a running change. Doing so leaves the development team free to concentrate fully on the agreed-upon goals so that they can meet the schedule and start on the next project soon.

Obviously, freezing a design requires a great deal of discipline on both sides. It can be a difficult behavioral transition for a company to make, but it is a critical one if the company is to cut its development cycle. It is also a critical issue for management, because it is through such changes that management signals to all its seriousness about shortening the development cycle. If management violates, or allows to be violated, a freeze on a design, skeptics will be quick to respond with "See, I told you so; this is just another passing idea of the boss's." (See Chapter 14 for more on management's signals.)

Have a Long-Term Plan

A company that creates a rapid development process and learns how to innovate in small steps is in the position of a person who buys a high-performance motorcycle. The motorcycle can get you to your destination faster, while increasing your sense of pleasure, but it can also get you into difficulty faster. Increasing the responsiveness of the development process is beneficial, but to exploit this speed a company needs supporting strengths in product planning.

These strengths are not so much an addition to a firm's marketing capabilities as a change in emphasis. Incremental innovation provides a responsive means of keeping in touch with the marketplace that traditional market research cannot match. There is less need for long-range forecasts and intensive market research to determine product features when an organization can move in quick steps, because much of this information is available just by watching which products sell and which don't. Offsetting this simplification is the greater need to have a product strategy capable of orchestrating all the small product steps.

Without a well-enunciated product strategy, a company that is responsive to the marketplace is likely just to get led around by the market. It will follow the hot-selling trends for a while, but its product line will become increasingly fragmented and the products may drift away from the area of the company's strengths. Besides a responsiveness to customers' immediate changes in preference, the product line needs focus and a theme that will keep it in the mainstream over the long haul.

For example, 3M has a well-known ability to introduce new products continuously and maintain a 25-percent level of products that have been introduced within the past five years. But 3M also has a deeply ingrained concept as to what constitutes a 3M product. Although their people are always working on a broad variety of product ideas, these all fit into a commonly understood framework.

A product strategy should state what is driving the product line, whether technical expertise, manufacturing capability, distribution channel strengths, or a certain type of customer base. The strategy should ask what types of products the company wants to embrace or avoid, as, for example, products under heavy government regulatory control or ones that conserve natural resources. It should take into consideration the company's view of the future and how it wants to participate in that future.

Once there is a product strategy, remember to communicate it to everyone involved in developing new products. Too often such strategies are developed only as an exercise by senior management or for just the top brass at headquarters and are never used or communicated to the people who must make the daily decisions. This information needs to be known by both the woman preparing the capital authorization for a new product and the man checking out a prototype in the lab.

After a company has a long-term product plan that everyone understands, it can be responsive to the market without being flustered by it.

Step on Your Own Toes—Occasionally

Those who dance in quick, short steps occasionally step on their own toes, and companies that relentlessly introduce new product variations will occasionally take some sales from their own products.

If we had perfect information about our competitors' upcoming new products, our own product development could be done much more effectively. In particular, we might not introduce an improvement to one of our own products until seeing that a competitor might encroach on our sales. Our market intelligence is seldom so accurate, however, so we have to make some assumptions about what competitors might do.

Companies that move quickly in the marketplace usually assume that if they can develop a product, so can a competitor. They know they may be the best, but only to a degree. If they delay in developing a new product idea, a competitor will most likely beat them to market.

This leadership strategy wins support when a product idea is something distinctive that will expand the product line. But what if it will undercut sales of a product introduced six months ago, the one with the large tooling budget that has not yet been amortized? This difficult issue was addressed thirty years ago by Ted Levitt in his classic article "Marketing Myopia" (see Suggested Reading at the end of this chapter). Either you put your own products out of business or your competitors will do it for you.

Let's consider the seemingly reckless tactic of cannibalizing one's own products. If the company goes ahead with development and a competitor is indeed active in this area, it will be a wise choice because the company will have stayed ahead of the competitor. But even if the competitor is not active, the company will still gain substantially. It will now be two steps ahead of the competition, with a solidified lead. Now the competition will be forced into either dropping out or attempting a highly risky leapfrogging operation. Although sales of the recently introduced product may be sacrificed, the company will probably have gained far more strategically in strengthening its hold on the market segment.

This is not to say that good product planners are always looking for opportunities to sacrifice their best products. Sales from good products do pay the bills and earn profits for the company. It is important, however, not to be petrified by fear that a new product may cut into sales of a current one. If planners were always looking to see what might disrupt current sales, nothing would get started for fear of upsetting the status quo.

Our advice is to look forward, not back. Ask how a particular product will position the company for the future, not how it might disrupt what was built in the past. By wearing forward-looking blinders you may occasionally cannibalize one of your own products, but that is a price that a company practicing incremental innovation can afford to pay.

SUGGESTED READING

Hagel III, John. 1988. Managing complexity. *The McKinsey Quarterly* (Spring): 2–23. An illuminating discussion of the impact of product complexity on development difficulty, particularly as reflected in development expense. Includes examples from the electronics industry.

Levitt, Theodore. 1960. Marketing myopia. *Harvard Business Review* 38(4): 45–56. A popular piece illustrating how companies dry up because they view their businesses too narrowly. It argues that incremental innovation can help companies remain in touch with change. Continually offer customers a fresh assortment of products, then watch how they define your business.

Levitt, Theodore. 1966. Innovative imitation. *Harvard Business Review* 44(5): 63–70. Levitt argues that breakthrough innovation receives more attention in the press than it deserves. The greatest flow of newness comes from imitation, not innovation, and most progress results from imitating something that already exists. This being the case, the wise company will concentrate on being able to imitate quickly.

CHAPTER 5

Product Specifications

A specification is a written description of a product that is generated beforehand to guide the development of the product. It often goes by such other names as a product brief, design guide, or product action form.

Specifications provide input that is critical to both the product and the development process. A poorly executed specification can both delay the start of development and stretch out the development cycle. The techniques covered in this chapter will help avoid writing specifications that might prove slow to win approval, yield overly complex products, waste resources, and thus impede product development.

WHY SPECIFICATIONS MATTER

Most product developers churn out specifications as though they were just another necessary piece of paperwork. They are most concerned to get it written, have it revised and approved as needed, then get on with the real job, which is developing the product. The specification's main use thereafter for most developers is to prove one's innocence in case things don't turn out as desired.

However, the specification is only the jumping-off point for the product; everything downstream will depend on the nature of the initial instructions. Here, for example, is where product complexity is determined, though often unknowingly. Any features or nuances added later to the specification may seem innocent enough in themselves, but they can add complexity, and thus time to the cycle. The compound relationship between complexity and development effort discussed in Chapter 4 means that a small addition in complexity can have a large impact on the schedule. Normally a schedule is thought of as a separate chart where tasks are laid out across a time scale. But this schedule really depends on the amount of work to be done, which is determined in the specification through the choices made there regarding product complexity.

Small additions to a product's features do not always cause big delays, but they may. The specification-writing stage is where this impact can and should be determined. By involving the right parties in the deliberations about specifications the impacts arise, are understood, and then resolved. If there is no resolution of the problem of impacts at this stage, the project schedule will remain fiction, however elaborately it may be drawn on a wall chart. The specification is typically the foundation of all project planning.

The specification is also a powerful tool for building a sense of ownership in the product's development. A person who has been involved in deliberations about a particular point and said, "Yes, I think we can make assembly A work with assembly B" has effectively signed up to make it work. If that same person has not been involved, he or she will likely find many reasons later why A and B won't work together.

It is natural to ask whether time spent in wrestling with the tradeoffs inherent in a product and ensuring individuals' involvement simply delays the program. In fact, many companies seem to try to get the specification stage over with quickly by avoiding as much controversy and involvement as possible. In contrast, let us suggest that this task be given genuine high-level attention. Later in this chapter we show how specifications can be written both better and faster than they are now in most companies.

Examples of Weak Specifications

A number of pitfalls and organizational obstacles make it difficult to generate an effective specification, and the basic concept of describing a nonexistent physical object in words and numbers has its own inherent difficulties. Three examples illustrate the kinds of specification-based problems that can cause products to miss their market potential and/or be late to market.

The first example involves an electronic instrument for the consumer market. The proposed product is an outgrowth of the company's basic business, but it involves a new type of sensing technology and a somewhat new market segment. (We have normally found the combination of new technology and a new market segment to be fraught with the potential for failure.) Marketing wrote the specification and provided what was easiest for it to provide—typical boilerplate information on product size, weight, power consumption, modes of operation, price, and sales volume. The critical issue from the start was measurement accuracy, but interestingly the specification gave no guidance or performance targets on this key topic.

Engineering tested the available technologies and competing products, which presumably were selling acceptably well, and reported that the measurement approach was feasible, though its accuracy was limited. Meanwhile, the specification's proposed product introduction date came and went with no progress being made on the product. Marketing did not know the market value of accurate measurement and how far the product could fall short of perfection. Engineering had not assessed the cost of improved measurement accuracy by testing variations in technology or design to improve it. Beyond these two departments, the company as a whole had yet to wrestle with the issues of quality image or product liability to know what was acceptable. Competitors were apparently making money on this product by learning how to satisfy customers on the accuracy issue by getting their feedback, and were strengthening their market position. While competitors went on entrenching their positions, this company continued to fall behind because it was not dealing with the key specification issue.

The second example involves a piece of industrial hardware. Marketing, working in isolation, specified that the product should be a first quality, top-of-the-line product but have the manufacturing cost of a mid-line model. Moreover, the spec called for the product to be developed in less than a year, in contrast with the typical development period of two to three years for that type of product.

When engineering received the specification, they just laughed. To them it was clearly an inconsistent set of goals. (Whether the goals were indeed inconsistent is debatable; it matters only that engineering perceived them to be, for they then washed their hands of responsibility for the outcome.) Rather than debating the product goals with marketing, an argument they thought themselves likely to lose, engineering decided on the basis of their own rather slanted view of the marketplace to design the product they had always wanted to design anyway.

The result was certainly a top-of-the-line product, elegant in anyone's opinion and a winner of design awards. Its cost, however, was well over twice the original target value. Even by cutting its profit to zero the company could not price the product low enough to attract many sales. It thus was an aesthetic masterpiece but a commercial failure. Moreover, the organization spent so long debating and trying to reduce product cost that in the end this product became one of their slowest development programs.

The third example is a food-processing machine. As marketing and engineering were to find out rather late in the development cycle, this machine's processing effectiveness was directly related to its height, and the quality of the processed output was of prime importance. However, the specification gave considerable latitude on height, saying only that the machine should be as low as possible and not exceed eleven feet in height. Quite liberal guidelines were provided also on food-processing quality. Marketing produced a four-foot-high mock-up of the machine. Lacking more definite height information or a clear statement on output quality, engineering worked from this image of a four-foot-high model.

Engineering finally got a seven-foot-high prototype working with fair effectiveness and thought they were doing badly because their prototype was higher than marketing's model. The real problem, they found out when marketing sampled the output, was that it did not process the food as completely as necessary. Marketing would have been quite happy with a ten-foot-high machine that processed food more thoroughly, but the relative importance of the specification's conflicting goals was not discussed at the outset. Because of this weak specification, months of extra time were wasted by focusing on the wrong design goals, and more months were needed at the prototype-testing stage to define the quality of the output and redesign the machine to produce acceptable output. These months proved costly for a product that originally had had only a one-year development schedule and a waiting market.

Why Specifications Are Hard to Write

As these examples suggest, inadequate specifications can easily misdirect design resources and waste time. But it is difficult to write specifications that map out unexplored territory. One major difficulty is that many types of knowledge—engineering, materials, marketing, manufacturing, and others—are needed to define the product, just as they are to develop it. This means that many specialists must thus provide input to the specification; it cannot be written by one person. Even having one person gather information from many others is not satisfactory, because all the specialists have to interact to share information and reach a mutually satisfactory solution.

Some of the most crucial elements in a specification are often hard to put into meaningful words or quantify. In the examples above, the factors of measurement accuracy (in this particular case because there were many uncontrollable variables), "top-of-the-line quality," and processed-food quality, respectively, were all difficult to specify in advance. Qualities like having "modern, high-tech styling" or being "user friendly" can have a great influence on product sales, but they resist accurate verbal description.

Even when a product's basic attributes can be defined, difficulties will arise because of various elements' interactions. At its heart, the design process requires the common solution of conflicting requirements by making trade-offs. For instance, a machine will process food better if it is taller. How tall should (not *can*) the machine be, and how well should the food be processed? The best design is therefore probably not the single point suggested in the specification but rather some other combination of values that suits the user better. The designer needs to have information that is in effect written between the lines of the specification in order to make appropriate trade-off decisions. Even a well-written specification will not contain between-the-lines information, so it will be incomplete by itself.

The last difficulty involved in writing a specification is that some of the information that should go into it does not become available until the team gets into design. Design is a learning process, some of which may show that the specification was shortsighted. Trade-off possibilities, as depicted in Figure 5-1, are one type of information coming out of the design activity that would alter a specification were it known beforehand.

Organizational Dynamics Thwart the Specification Process

Chapter 4 provided descriptions of engineering, marketing, and manufacturing goals to show how each function's goals encourage long development cycles. Although these goals work in unison to slow development, they also present conflicting objectives when trying to decide on product attributes. Consider the differences.

Marketing, because it is strongly allied with sales, tends to think in terms of product features. If, for example, a competitor starts offering touch-screen technology in its computer terminals, we must add a touch-screen capability to our units so that customers will not "disqualify" us for not having this feature. Whether the feature is actually beneficial is of secondary importance. If specifications were written solely by marketing, as is too often the case, they would tend to emphasize features, which would be selected in relation to those of competing products.

Engineering's bias is to experiment with new technology because

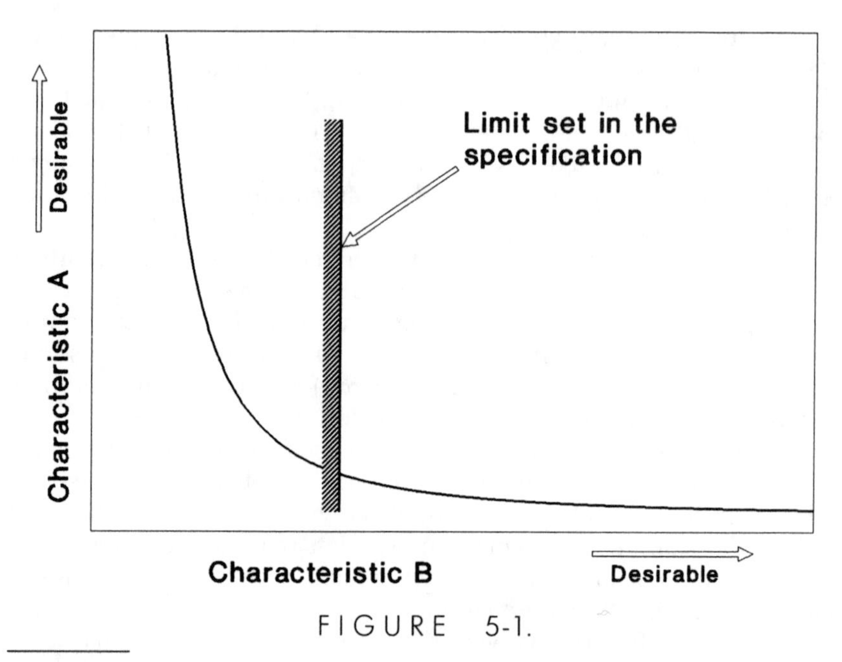

FIGURE 5-1.

Because product characteristics interact in perhaps unknown ways, limiting Characteristic B in a specification may force Characteristic A into a region that provides poor customer value.

it is exciting and there are always fresh discoveries to be made. Many technologies change quickly, and engineers enjoy the challenge of gaining experience with new technologies.

Manufacturing operations thrive in a stable environment. For them bringing a new product into production means starting at the top of the learning curve. Manufacturing people are not rewarded for the headaches they incur by being involved in working the bugs out of a new product. Pilot runs of new products tie up valuable assembly lines, new types of parts complicate inventory problems, and new manufacturing processes occupy both floor space and the time of their most valuable troubleshooters. The employment of people in manufacturing may depend on new products in the long term, but there is no immediate payoff in new products for manufacturing.

These sketches are stereotypes, of course, but they do illustrate that different functions approach a new product with their own unique objectives. We believe that having a well-written specification is the key to getting off to a fast start on a new product and that specs can be written well only by combining the knowledge of different functions. But how can this be done if the functions are all pulling in different directions?

Marketing Specs versus Engineering Specs

Before we leave the problem of writing specifications and proceed to the solutions, there is one more point to cover. Some organizations recognize that marketing and engineering write specifications for different purposes, a disparity they deal with simply by having two specifications. This practice is especially prevalent in the electronics, particularly the computer, industry. There, marketing writes its specification from a sales and business perspective, unencumbered by technical jargon. Then engineering writes a specification that really describes how the product is to work in its most minute details.

We advise against adopting this practice because having separate marketing and engineering specifications is dangerous in two respects. First, it cleverly skirts the essential problem of getting marketing and engineering to work out their differences and reach a common view of the product. Engineering thus never has to understand rather subjective market research suggesting why customers want certain features, and marketing avoids having to work through esoteric engineering parlance to figure out the pros and cons of a technical approach. The organization thus never gains the benefit of making sound trade-offs between features and technology.

Second, we have consistently found that when there are two specifications they describe two different products, even if one originally derived from the other. As each specification evolves, it changes and takes on a life of its own. Engineering forgets that there is a marketing spec and regards its own as the only guidebook. Marketing frequently can't understand engineering's extremely detailed specification and thus ignores it.

Only one document in any given organization should be called the specification, and it should incorporate the joint labor of all parties. What is usually referred to as the engineering specification should be considered an engineering document that is useful to initiate design but is emphatically not the definition of the product.

WRITING SPECIFICATIONS FOR SPEEDY DEVELOPMENT

The document designated as an engineering specification is actually the antithesis of a specification because it is far too narrow and detailed. It deals with implementation details and answers questions dealing with how rather than what.

Creating a specification should begin by focusing on the customer. Who is the target customer? What is that customer's problem? How will the product solve this problem? What benefits will it offer as compared to other solutions that are available to the customer? How will the customer value these benefits? On what basis will the customer judge the product?

Everyone involved with a development project must understand who the customer is expected to be and what needs will be satisfied. This understanding is essential, but not easily arrived at. Most specifications provide merely a perfunctory one- or two-sentence statement of the raison d'être for the product. A much more detailed discussion is needed, though, to help those who will be making trade-off decisions later so that their decisions fit in with the purpose of the product. A solid statement on the product's mission focuses on the customer and the need for the product, and guides developers toward answers for all the questions that are not explicitly covered by the spec.

To appreciate why a broad understanding of the specification is needed, look back at the three examples of weak specifications given earlier in this chapter. In the case of the electronic instrument, this company got bogged down because of not knowing who would be taking readings with the instrument, how accurate they expected or needed the readings to be, or how pleased users were with competing instruments. They knew how to make the instrument but just didn't know how good it had to be, which stymied them. In the industrial hardware example, the crux of their difficulties and the resulting delay was that the designer had no idea how customers would value the product he was designing, so he gladly filled in the information gaps with his own hunches. For the food-processing machine, little advance thought had been given to the fundamental question of how well it had to do its job or what the relative values to the customer were of its effectiveness and compactness.

In all three cases the designers, the people actually creating the product, did not firmly focus on a real customer or that customer's needs. Having a solid statement on product intent that describes the customer and that customer's needs is essential to keep the design team headed roughly in the proper direction.

Involve Customers and Users

A specification has to relate a product to the marketplace. There is no better way of doing this than by actually involving real customers or users in writing the specification. First let us see how to find appropriate users, then show how they can be employed.

The most helpful users are usually the lead users, the innovators, who might be the first to use a new product. They are the most demanding and most aware of the shortcomings in competing solutions now on the market. Lead users are found by first identifying important, new, or leading trends in the marketplace in question. If the company has a strategic plan it may provide this information. Identifying these trends is easier in markets such as most industrial ones that follow some rational pattern and can be predicted to some extent by using economic or technological trends.

After identifying the trends for a given business, look for user groups that exemplify or are most likely to be on the leading edge of the trend. Say, for example, that you manufacture telephone sets. Recognizing that the population is getting older, you would like to target an apparent growth market in telephones for older people. Existing phones may present problems to those with hearing, visual, or dexterity limitations. Lead users might be sought out in nursing homes, hospitals, or senior centers or through visiting nurses' associations or retiree groups. Let's say you also want to identify the lead users who are most likely to be innovative. Out of the groups just mentioned, hospitals might be the best choice because they are used to buying new high-tech equipment and trying it, whereas the other groups mentioned might not give hardware much thought.

Lead users often are ones who will get the greatest value from the innovation and be willing to pay a premium price for a product that meets their needs better than other items now on the market. These users may be both dissatisfied and prepared to buy their way out of this dissatisfaction. These users may, however, also be in a market that is different and more advanced than the market you have chosen to serve. For instance, hospital telephones may have hygienic requirements that are unnecessary for use as the residential phones that may be your market of main interest. Consequently, you may have to employ the lead users carefully, recognizing that there are some places where they can lend insight but some where they could lead you astray.

As just suggested, lead users can be dangerous if relied on too heavily. A true lead user will be someone at the pinnacle of a pyramid, but you will also want to make sure to satisfy others further down the pyramid. In designing car radios, for example, one might find a lead user who drives a top-of-the-line BMW. He or she might have bought this car partly because of wanting to impress passengers with the technical wizardry on the dashboard. Offsetting such a lead user is the technophobe who avoids buying cars with complicated radios. This user, who is perhaps more common, needs also to be considered.

The stress is on users rather than customers here, because when designing or specifying a product users generally provide the most pertinent information. Consider, for instance, an office word processing machine. A person who sits at the keyboard forty hours a week will be in the best position to explain the machine's shortcomings. The employer who paid for it or the dealer who supplied it will have different—but normally less valuable—information on the product's attributes and weaknesses. The best approach usually is to work closely with users, then check with other buying-decision makers and those in the distribution channel to make sure you have not erected any obstacles for them.

Many companies are reluctant to involve outsiders in the process of designing or specifying their products. There are such questions to be considered as how to protect proprietary information and to compensate outsiders. These difficulties are rarely insurmountable once a company recognizes the value of incorporating user information. If this is a big problem, insiders can often provide a great deal of user information. When writing a specification, consider bringing in field-service technicians or the person in the laboratory who tests your own or competitors' products.

Once there are users to help, several ways to employ them can be set up. They can be used directly as participants in specification workshops, described below. It is also useful just to keep lead users in mind as the people for whom you have to design. Get to know a few of them and write their phone numbers right on the specification as reminders of the audience for whom the product is being specified. Once the phone numbers are on the specification, call them once in a while. Use them as reality checks on the work to make sure you are developing a product for real customers.

Focus on Product Benefits, Not Just Features

Once a product's intent is clear, the next step is to define its benefits. Too often, specification writers work with features rather than benefits—but customers buy benefits, not features. For example, in writing a specification for a fax machine it may be tempting to specify a stepper motor or servo drive to move the paper or print head. But is that a benefit? Can the customer perceive it directly? If not, then we should specify what the customer will actually notice, such as speed, print quality, or noise level.

Two traps result from concentrating on features. One is that certain ones may be unnecessary to achieve product benefits, and imposing them only adds unnecessary design constraints, as in the fax machine example. In this case the final choice of motor technology should depend on size, cost, or power restrictions that appear as the design materializes. When a designer concentrates on benefits, he or she can select from a richer, more value driven set of features and may be able to eliminate the need for a particular feature. For example, early FM tuners had an AFC (automatic frequency control) switch that became a valued feature because it kept the tuner from drifting off-station. Then as designs shifted to more stable phase-locked loop circuits, this feature became superfluous.

The other trap associated with concentrating on features is that they can hide a "fatal flaw" problem, which occurs when several features must work together to provide a benefit. If one key feature, the fatal flaw, is inadvertently omitted, the entire benefit will be lost. In one such case a company designing a computer printer for office use neglected to provide for compatibility with the two dominant suppliers of office automation equipment. The development team ended up worrying more about print speed and print quality than about where customers might use the printer. It is easy to blame the designer for not including the missing feature, but the oversight really stems from focusing on features rather than on the benefits that derive from a cluster of features.

Do Not Restrict Design Options

A specification should concentrate on what is needed in its product, not on how to do it. There are often several technical options available to designers to achieve the stated design objectives. Designers should be left free to choose among the various implementation options as the design materializes, exposing new constraints or trade-offs. This flexibility will allow designers to make choices that can save time. For example, one design avenue might entail a long lead time in tooling, but another might use an available part instead.

Those who write specifications usually know that they should avoid specifying the means of implementation, but they often do so anyway, because it is easier to list "316 stainless steel" than to give the properties needed for an acceptable product. What is often not appreciated, however, is that the designer must be intimately involved in setting product objectives in order to have the background needed for making the best trade-off decisions. If designers do not understand the game plan, their personal preferences will prevail.

Involve All Functions in the Trade-Offs

At a minimum, engineering, marketing, and manufacturing should be involved as a team in writing the specification. And, depending on the product and how the company works, other functions such as purchasing, quality assurance, testing, or customer service may be brought in as well.

Involvement means joint participation in actually creating the specification on an equal basis, not playing Ping-Pong by sending the spec back and forth between departments. In many companies one particular department, usually either engineering or marketing, creates the specification, then passes it on to other functions for review, comment, or approval. When there is little response, they will assume that everything is all right, everyone agrees with their view of the product, and they can proceed with the project. More likely than not, however, no response means that people have yet to read it seriously, do not understand it, or simply don't think it important enough to merit a reaction. Under such conditions the originator will start off by thinking that the product idea has been sold and everyone is excited about it, but in fact there is no enthusiasm or momentum. The new product is thus not getting the strong thrust into development that is essential to its overcoming corporate inertia.

All functions need to be represented in the process at the outset because each one has a unique type of expertise to contribute. In principle, this expertise could be injected during the specification review or approval process, but unfortunately the path is not so straightforward. The originator, with no devious intent, may well formulate the product in such a way as to cloud the areas needing input from others. Then, once the specification has been written down, the alternative formulations or underlying assumptions will no longer be apparent. The creating of a specification is simply too fragile a process to be done sequentially.

As discussed earlier, the essence of a good specification is balance among the many design trade-offs that must be made as the product goes into development. Joint participation in writing the specification gets these trade-offs out in the open and starts the extensive interplay needed to resolve them in a way that will yield a competitive product. The trade-offs cannot all be resolved in writing the specification until some design work is done to provide more information. But if the key players can be involved in the initial discussions they will be able to reach consensus on the direction of the product and understand the rationale for the fundamental product decisions. They will then be capable of making consistent trade-off decisions later in the process.

The other, perhaps more important, outcome from involving all key players in writing the specification is that in the process of contributing to it they in effect "sign up" for the project. This commitment or buy-in is essential to getting the project completed quickly. Without it, people brought in later will spend time learning about the project, wondering why it is not being done differently, and generally just picking it apart to see if they like it. When a person actually contributes to writing the specification, though, they are saying, "This can be done. I agree with this approach, and I will find a way to make it work." Because it is embarrassing to the writer when a specification turns out to be infeasible, the writer is likely to put effort into making it successful.

Identify and Deal with Crucial Factors

Most specs are too long, but even with all their verbiage still do not describe the most important aspects of a product. People tend to write about what they can describe, things they understand. The factors most likely to delay a product are the ones nobody understands, which seldom get written into the spec. Specification writers must therefore force themselves to identify and describe the crucial elements of the product, because the easy ones will take care of themselves.

Avoid putting trivia in a spec, by not using a standard boilerplate product specification form. An engineer in a company that makes household appliances once told us, obviously feeling insulted, that "They [marketing] specify that the product has to be UL listed. I *know* that; it's company policy." The point is that a product specification is not a legal contract. If the product does not turn out as expected, there is no way to sue somebody else in the company to recover damages anyway, so don't bother with this kind of protection.

Instead, build some custom checklists that will concentrate attention on areas where the specifications have failed before. Do not, for instance, use material supplied by 3M or Hewlett-Packard, as enviable as their innovation records may be. Using other companies' checklists will just add irrelevant, time-wasting steps. Go through some completed projects and identify product-definition problems that have snared your staff before. Try to generalize these problems a bit and use them to create the custom checklist. Here are some broad categories to get you started thinking, but do not use these as your checklist:

- Interactions with associated products, past, present, or contemplated
- The potential for design growth or modification
- The physical environment in which the product will be used

- Patent infringement/protection
- Safety and liability
- Quality and reliability
- Ergonomics
- The users' abilities
- Sourcing and assembly
- Distribution
- Documentation, training, servicing, and maintenance
- Unusual equipment or facilities needed

Another way to develop a checklist is to put yourself in the place of the customer or user. Ask yourself a variety of questions designed to illuminate issues that have held up previous projects: What other alternatives are available to customers? Why might they choose another product instead of this one? Is this product going to save customers time, money, or aggravation? If so, how? What difficulties might a customer encounter in adding this product to the setup they already have?

Specifications do not have to be long. In fact, every page added to a spec increases the chances that people will not read it. Companies like Hewett-Packard and Compaq, which develop sophisticated products on short schedules, often write a specification on one page. Is this a reasonable goal for your own products?

HOW TO WRITE SPECIFICATIONS JOINTLY

There is a way to write specifications so that they achieve the beneficial objectives discussed above. Again, the objectives are to write and approve a spec quickly while encouraging the key participants in the development effort to buy into the project. It is also important to include the diverse information that is needed from users and various disciplines so that the key design trade-offs are exposed. Finally, we need to think broadly in order to concentrate on the key design factors.

In our experience these goals can all be achieved most easily by using an off-site workshop in which the key participants create the specification jointly, starting from a broad statement of the need for the product. It is important that the workshop begin with a fresh start on the specification. No function should arrive with an initial product definition expecting that the other participants will use it as a starting point. Everything about the product is to be decided in the workshop, and everyone is to leave there committed to what has been decided by the group. The purpose of the workshop is not only to produce a specification but also to build support for, and an understanding about, the product among those who will have to bring it into being.

Clearly, the workshop participants should be chosen carefully to ensure that there is appropriate representation among those who will have to execute the resulting specification. This is a powerful opportunity to build project buy-in, and it should not be wasted. Remember to include users or user representatives also. The workshop is for doers, not top management. Managers should participate only if they can attend the entire workshop and only if they will be involved in the project on a daily basis. Otherwise, they may attend as silent observers. There is, however, one exception: one senior manager should kick off the workshop as a keynote speaker, indicating in broad terms how the product fits into management's vision of the company's future. Workshop participants then need to assimilate this vision as a context for defining the product.

The workshop should be led by an individual with no stake in its outcome. Most often this will be an independent, outside facilitator. The principal project participants, such as the project leader, head engineer, or marketing product manager, are particularly inappropriate to serve as workshop leader because of their vested interest in its outcome. They must play a role by advocating specific viewpoints, which undermines their ability to be an objective moderator. Furthermore, their contribution to the content of the specification is too important for them to be distracted by trying to keep the meeting running smoothly.

The facilitator is responsible only for the process of the meeting, which becomes a demanding, full-time job when conflicting interests arise, as they will when alternatives and trade-offs are exposed. The content of the specification is the responsibility of the participants, not the meeting's facilitator. A neutral facilitator can often be obtained from a department having little to do with product development, such as human resources. Training and development people are likely to be trained in how to lead meetings and deal with conflict. Another division of the company or its headquarters office may be other sources of facilitators. And some consultants specialize in this type of work, being particularly adept at leading participants in exposing and resolving key product issues.

A workshop should be held off-site, to minimize distractions. It typically takes about three days. The participants should be briefed beforehand to explain the objective and the rules. Start the workshop by discussing broadly the need for the product and the benefits it should provide. Work down to details like product features, functions, performance, and cost. Recording the proceedings on flip charts that then can be taped on the walls makes it relatively easy after the workshop to transform them into a document.

Quality Function Deployment

Some companies deal with specification issues by using a somewhat similar process called quality function deployment (QFD). The outcome of QFD is not a product specification in the conventional sense, but the QFD process does provide another vantage point for viewing the product concept that exposes its critical linkages. The similarity with the workshop technique just described is that both methods force people from all the functions involved to identify and resolve the key issues in a product concept early on, in a relatively structured, unbiased format.

The QFD process is built around a sequence of matrices that relate various quantities. These matrices transform the product attributes desired into design parameters and finally into production requirements. The clear linkage between product benefit and manufacturing variable is valuable for tying the various marketing, engineering, and manufacturing interests together and taking the voice of the customer into account. These linkages can also expose such key product design parameters as the height of the food-processing machine cited earlier. In addition, QFD can compare the design parameters for a proposed design with those for competitive designs by employing data obtained from competitive analysis techniques like benchmarking.

QFD can effectively complement the specification workshop technique. As it is normally used, QFD cannot develop a product definition as quickly as can a specification workshop. But it could supply a base of supporting data for a specification workshop, or be applied after the workshop to project the specification parameters on into the manufacturing process.

The QFD process can be mechanical and tedious if users get carried away with filling in all the boxes and lose sight of the big picture. There have been complaints that by the time a QFD analysis was complete the design had progressed beyond the point of being able to use its insights. QFD seems to work best for attributes that are easily quantified and can be linked to one or just a few design parameters, so there are situations where it may not be well suited to generating specifications. Like many other analytical techniques for capturing customers' desires, QFD appears to work best on established products to which people can easily relate, in contrast to a concept for a product that does not exist yet. This may explain why QFD has had its strongest reception to date in the automobile industry. In many applications, however, it provides a means of capturing linkages that would otherwise fail to be recognized and for emphasizing critical manufacturing parameters much sooner than would otherwise happen.

QFD is just starting to be used in North American industry, with leading companies beginning to experiment with it. We have no first-hand experience in applying it to a product design. Because our policy in this book is to recommend only those techniques with which we have had direct experience, we cannot provide a recommendation regarding QFD at this time.

Dealing with Specification Changes

The specification is by nature an imperfect representation of an imperfect product. As a team learns how to get off to a faster start, it will be starting with less information, so that the spec will be more likely to have shortcomings. This is why it is so important for everyone on the team to have a broad view of the product, its goals, and its benefits. A good development team can usually fill in missing details or make adjustments if they have the big picture clearly in mind.

Recognize that the specification is just a means to an end. Do not put off writing the spec because of not having all the information required—you will never have it all. You will need the specification on Day One, so write up what you have at the outset and keep track of what needs to be added later.

It is harder to write a specification while you are trying to get a fast start on a project, but accelerated product development also eases some of the need to modify specifications. The moving-target concept (see Chapter 3) suggests that product objectives are always in flux anyway, so the longer we take to develop a product the more likely it is that we will have to change the fundamental product objective. It is easier to change a few details we may not be quite sure about than to have the basic product goal change in the middle of development. Incremental innovation strategies (see Chapter 4) also help to soften the effect of any inadequacies in the specification. If you plan to develop another generation of the product soon, many shortcomings can then be corrected in it without delaying development of the current generation. The ability to develop a product quickly provides a means of coping with various shortcomings in the product's attributes.

SUGGESTED READING

Doyle, Michael, and David Strauss. 1976. *How to Make Meetings Work*. New York: Jove Books. Specific skills are needed to run facilitated meetings like specification workshops. Although the prevailing techniques have improved since this work was published, this classic describes the basics well.

Fisher, Roger, and William Ury. 1981. *Getting to Yes.* New York: Penguin. Reaching agreement on a specification by several parties is essentially a process of negotiation, and skill in negotiating definitely helps speed up the process. We like Fisher and Ury's approach because it parallels our own of working from product benefits to features. These authors stress that it is more productive to understand each party's underlying interests than just to state positions, which then tend to become inflexible.

Hauser, John R., and Don Clausing. 1988. The house of quality. *Harvard Business Review* 66(3): 63–73. Probably the most readable description of QFD generally available. Examples show, through a series of matrices, how customer attributes lead to production requirements.

King, Bob. *Better Designs in Half the Time: Implementing Quality Function Deployment in America.* 1989. 3rd ed. Methuen, Mass.: GOAL/QPC. This self-described cookbook provides relatively little information either on designing products or on cutting development time. It does, however, provide instructions for building dozens of QFD matrices to relate the "voice of the customer" to critical design parameters, then to production variables. In so doing it should encourage the various parties concerned to identify critical trade-off issues.

CHAPTER 6

The Subtle Role of Product Architecture

After the specification process we enter another key stage in design, one that most companies unconsciously try to avoid. This is the system design phase, which sets the fundamental architecture of the product.

"Architecture" is indeed an appropriate term for it. To use the analogy of building a house, this is not the stage of selecting the color of the paint, the size of the cabinets, and the carpets. Rather it is the stage in which to decide how many rooms there will be, how big they will be, where the walls are to go, where the doors and windows go, and how high the ceilings should be. In other words, the decisions made now almost irreversibly control what can be done later in the design process. It would be very difficult to put in nine-foot-high cabinets if the ceiling is eight feet high.

Experience suggests that the success of a product is determined largely by the decisions made in the brief system design phase. If it is done well, it creates the perfect foundation upon which to build the rest of the design. However, if it is done poorly there will be almost no chance of recovery. More to the point here is that certain architectural choices provide more opportunities than others for using time-saving techniques such as concurrent activities.

Too often, key architectural decisions are made almost automatically. These decisions are most commonly driven by a desire to achieve product performance goals or reduce product costs. Too few companies see architecture as a tool for achieving rapid product development.

They frequently rush through the architectural stage to get to detailed design activities.

The interesting paradox is that the most highly leveraged portion of the design process is during the first 10 percent of it, not the last 10 percent. The first 10 percent is when the key system design decisions occur, which can determine 90 percent of a product's cost and performance.

This chapter focuses attention on this key stage and describes the principles of structuring a product for rapid development. Let us begin by discussing the key architectural decisions that must be made and how each of these is likely to affect development speed.

THE KEY ARCHITECTURAL DECISIONS

Here are the key architectural questions that have to be answered during the system design phase:

To what degree should functionality be modularized?
In which modules should functionality be placed?
How much reserve performance should be put in each module?
What type of interfaces should be used between modules?
How much technical risk should be taken in each module?

Each of these decisions has important implications for development speed because they will influence our freedom to manage the design process. It is easiest to see this with some practical examples.

CONCENTRATING FUNCTIONALITY

One key initial decision is whether the functionality of the product will be allocated among a number of specialized modules or left in a single central location. This can be viewed as a decision on the degree of modularity to incorporate in the product. This modularity has both advantages and disadvantages. In some cases it is preferable to use specialized modules within the design to accomplish specific tasks, in others to fit all the functions into one big module.

For example, in one case involving a cable TV decoder design a decision was made to locate the power supply in a separate box with a cable leading to the unit on top of the TV set. This added extra cost but provided several advantages: the unit on the set could be made smaller and more streamlined, it produced less heat and thus ran cooler, and it divided the design into two separate tasks.

The last point is the critical one with respect to development speed. The extent to which a system can be divided into well-defined chunks positions us to start work on those chunks independently. Working on these chunks simultaneously permits overlapping activities and reduces the length of the design process.

In our experience, effective use of modularity tends to shorten development cycles, but this benefit is certainly not free. Modularity is likely to impose two basic disadvantages. First, it adds cost because modules must be interfaced with one another. They may require mechanical interfaces such as mounting holes in a part that allow another part to be fastened to it. If the two parts could have been produced as a single piece, no interface would have been required, no holes made, and no fasteners or fastening operation required.

Modules may also require electronic interfaces such as cables and connectors. These extra parts add both material and labor costs to the product.

The second major disadvantage of modularity is that it can reduce performance. In many cases the interfaces between modules become the weak links in the system. Electronic circuits cannot escape the fact that electrons can travel only so fast. When functionality is located in distinct modules that are physically separated, the distances can slow down circuit performance. Elaborate, expensive technical tricks may be required to overcome this problem.

Similarly, mechanical interfaces can reduce performance. Think of the loose legs on an old wooden chair. The interfaces between the chair legs and the seat often prove weaker than either the legs or the seat itself. Many times an electronic unit can be "fixed" simply by wiggling a plug.

Normally, the effect of these cost and performance penalties will be insignificant compared to the benefits of cutting the time to market. In fact, over a full product life cycle the cost and performance penalties can turn out to be surprisingly small because of another interesting side benefit of modularity. Modularity makes it possible to upgrade overall functionality without completely redoing a design.

For example, consider a state-of-the-art oscilloscope. The technology of the input module will have a critical impact on overall performance. This technology will evolve faster than the technology in other sections of the product. In a case like this it is a good system design strategy to partition the subsystem that has the most rapidly changing technology into a separate module. As this technology evolves, only the module affected needs to be redesigned, and by planning in advance for the introduction of new higher performance technology the modular approach will aid rather than detract from overall system performance.

In much the same way, if a particular subsystem uses a technology that is evolving rapidly and dropping in cost, that subsystem should usually be modularized. Thus minimal redesign is required to capture the benefits of falling costs. This approach is used in personal computer designs for hard disk subsystems where the standard interface used on this module allows a system manufacturer to shift quickly to the latest disk drive or one that offers the best price–performance ratio.

Modularity can also be exploited to gain speed in another way. Often a new product will really be a whole line of new products with perhaps hundreds of variations to suit individual applications. By making some insightful architectural decisions a few core variations can be completed quickly, without compromising later expansion opportunities. Marketing people sometimes balk at this strategy because it is easier for them to sell a complete line, but with some open-minded and creative interplay it is often possible to find ways of getting started with lead users when only a few variations are available.

The applications for a modular line of products exist in many fields, especially in industrial or commercial goods. Let us illustrate this point by using such mechanical products as door locks or valves. These items come in various sizes, in left- and right-handed variants, standard and heavy duty grades, high and low temperature ranges, with various default or fail-safe options, and in a myriad of materials and finishes. The key to developing such a product line quickly is making a substantial upfront effort to plan an expandable product architecture that will allow for anticipated variations without having to design all of them initially.

In summary, degree of modularity is a key system design decision, and increased modularity generally enables subdividing the design task, permitting simultaneous work on different subsystems and a shorter development process.

LOCATING FUNCTIONS

Once the committment has been made to using modules, the next decision is which module to use for each function. This is a tricky, complex task. Some functions are naturally better separated and some are better off grouped. Good system design requires rethinking the way that functions are distributed among modules.

A great example of how to separate functions well was Hewlett-Packard's first low-cost plotter. Most plotter designs of the time used a mechanism to move the plotter pen in both the x and y directions. Because this mechanism was fairly massive, it required large motors to move it quickly over the paper. Hewlett-Packard recognized, however, that the function of movement did not necessarily have to be concentrated in the pen assembly. Instead they decided to move the paper in one axis and the pen in the other. Now the pen only had to move from left to right, while the paper moved up and down. Since the paper was light in weight it required smaller motors to move it, and it could move quickly. The overall result was a quick, inexpensive plotter.

Unfortunately, while this was excellent creative design work, it is not an approach that shortens development cycles. Shifting functionality from one module to another creates additional interface problems to solve. Creating extra problems to be solved lengthens the design cycle.

To keep the development cycle as short as possible, keep functionality in roughly the same module from one design to the next, and minimize movement of functionality between modules.

MODULE PERFORMANCE MARGINS

The third key system design decision is the amount of reserve performance to put into each module. Putting in too much costs money, and with too little the product won't perform adequately. Normally, we strive to put in just the right amount of reserve, and if necessary err slightly on the side of extra performance.

For example, a traditional design problem is designing a power supply for a piece of electronic equipment. The normal approach would be to design all the circuits that are to be fed from the supply, then design a supply that meets these requirements. Working sequentially avoids spending a penny more on power capacity than absolutely needed.

But does this make sense when trying to develop products quickly? Trying to put the perfect amount of performance in a subsystem requires understanding overall system needs exactly and having a precise understanding of all the subsystems with which it will interact.

This causes two problems. First, the more precise the design, the more engineering time it will take. Second, the information needed to do precise design does not become available until many of the other subsystems have been designed. The "dependent" subsystem must be designed after the ones that determine its performance level, which delays the design. Sometimes dependencies can become even more complex, when a design must be done interactively, jumping from module to module.

The overall effect of being stingy with performance is to slow the design process. Allowing generous design margins on all subsystems achieves a shorter development cycle.

In the example of the power supply it might be possible simply to estimate the power needs of all the circuits, add 20 percent as a safety factor, and begin designing immediately. With any luck the power supply would be designed even before the other modules were completed, and the risk of having to redesign the power supply module would be low.

Rapid development cycles are favored by having generous design margins in all the subsystems.

THE DESIGN OF INTERFACES

If modules are the bricks, then interfaces are the mortar that holds them together. Using the right approach to interface design can shorten development cycles. To do so means designing stable, robust, standard, and simple interfaces.

Having stable interfaces is vital because interfaces are the key external constraint of the module designer. If an interface changes, the module designer may need to redesign the module. Such a redesign could range from a relatively minor task to one in which the entire current design would have to be discarded.

Since the objective is to design as many modules concurrently as possible, interfaces must be defined early in the design process and prevented from changing. Doing so allows the module designer to work with a fixed set of rules.

For example, assume that we are designing a circuit that can expect to receive an input signal equal to one volt. If, halfway through the design, the module designer discovers that the input signal will be only one-tenth of a volt, another stage of amplification may be necessary. This new stage may introduce phase delays, a form of signal distortion. The overall result may be to create a "back to the drawing boards'' situation in a design that is almost complete. Having a stable interface minimizes the ''moving target'' problem for the module designer, thus improving efficiency and speed.

Let us not overlook that having a stable interface may well also improve morale. People will often give 110 percent when they are properly motivated. However, nothing destroys the motivation to give extra effort as much as seeing work that is almost completed rendered totally useless because the requirements have been changed. This is like trying to run a hundred-yard dash, only to be stopped ninety yards down the track and told that you are running in the wrong direction. It should be no surprise if after several such episodes designers simply slow down the pace. They will have learned that there is a lot less wasted effort by waiting for everybody else to finish before completing their module. Unfortunately, if no one wants to finish first no races will be won.

Thus, having stable interfaces is critical for both efficiency and motivation. The most powerful tool for making interfaces stay put is to make them robust.

Robust interfaces are desirable for the same reason as are generous margins in subsystems. An interface with excess capacity is unlikely to need changing. For example, it is not unusual in electronic systems to have subsystems called busses connecting modules. These are simply a collection of electrical conductors that hook one module to another. They can carry power as well as information between modules. In some cases a bus will be designed to work at a particular speed or frequency. Busses can be designed to work over a narrow range of frequencies or, with more difficulty, over a wide range. The wider the range of frequencies, the more robust the bus and the more flexible it is for the designer. Busses can be designed to have extra, unused conductors in them that may have to be used later if someone forgets to allow a path for a particular signal. This flexibility is what helps keep an interface stable as the design process progresses.

Mechanical interfaces can also be designed to be robust. If, for instance, a part is mounted to a foundation with larger bolts than required, then if the part becomes heavier it may be unnecessary to change the bolts. The same principles apply to both mechanical and electronic design. The more robust the interface, the less likely the need to have to change it.

The third principle of good interface selection is to use standard interfaces whenever possible. Standard ones have a number of important advantages: designers and suppliers already understand them, and their quirks will have already been discovered. However, they will not be perfectly suited for a given design. In fact, there is no standard interface for which an engineer of even average competence cannot identify at least three improvements needed for a particular application. A standard interface is in effect an "off the rack" suit, not a custom-tailored one. It will not fit perfectly but will usually do the job.

Because interface design is a difficult art, many improvements suggested to improve on standard interfaces turn out to be no improvement at all but simply a substitute set of problems. In our experience, brilliant designs do not often come from standard interfaces, but quick ones do.

The final principle in interface design is to keep them simple, primarily for simplified communication.

Any time there is more than one designer working on a product they will have to communicate with each other. With each additional foot of separation between designers, the communication problem becomes more severe. Each additional person on the design team increases the problem geometrically. Often the focus of this communication difficulty is on interfaces. "What does my module have to do for your module?" Making the subject matter simple greatly decreases the communication task.

Having simple interfaces is particularly important when there are participants in the design process from outside the immediate organization. It is not unusual for even the simplest messages to get confused as they travel from one organization to another. Different organizations may, for example, use the same specific technical terms to mean two different things. Keeping interfaces simple reduces the risk of communications problems.

DEGREE OF TECHNICAL RISK

In the system design process, a decision must be made on the degree of technical risk to be taken with each module. This step inevitably begins crossing the border into module design, at least to the extent that it formally or informally establishes the feasibility and risk levels of the technology to be used in each module.

There are two problems to manage. First, avoid asking for a "magic module" whose design requires magic powers to complete, violating the laws of physics or other improbable prerequisites. This problem is the one of having too much risk in a single module. Delays in completing the design of such a module will certainly hold back the entire product.

This problem can be controlled by establishing that there is at least one technically feasible way to achieve the module's function,

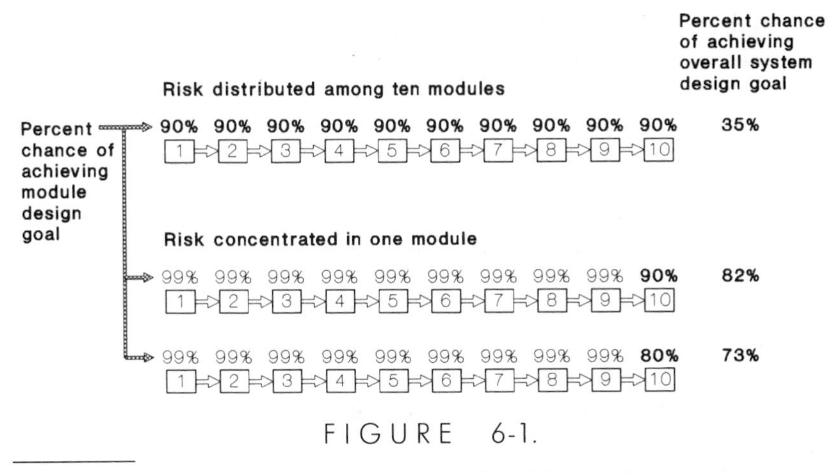

FIGURE 6-1.

If risk can be concentrated in a minimal number of modules, it will become easier to achieve system objectives.

then leaving it to the module designer either to implement this approach or find a better one.

The second problem is a more subtle one because it is risky without appearing to be. This occurs when moderate levels of risk are distributed to all the modules. The multiplicative effect of these risks makes the overall system's risk level excessively high. In our experience this is a far more common mistake than the first. The solution is to concentrate the technical risk within a few modules. Figure 6-1 illustrates the essence of this problem.

If there are ten modules in a system, each with a 90 percent chance of achieving its design goal, there is only about a one-third chance that the overall system will achieve its design goal. (The goal is a target cost, target performance, and a delivery date.)

Contrast this with the situation where the risk is concentrated in a single module. If nine modules each have a 99 percent chance of achieving their design goal and the tenth has a 90 percent chance, the overall probability of success would be over 80 percent. Thus, restricting the risky technology to a single subsystem more than doubles the probability of success.

In fact, by concentrating risky technology in a single subsystem we could use technology with an 80 percent chance of meeting its design goal and still more than double the chance of success from that in the first case with all modules at 90 percent.

There are several other benefits in employing proven technology for the majority of modules and concentrating risky technology within a few modules. Modules with risky technology tend to be communication intensive and their designers must communicate frequently with the designers of other modules. Concentrating the risk in a single area minimizes the amount of external communication, the risk associated with miscommunication, and forces communication to take place within a smaller, more cohesive work group. All these factors facilitate faster development.

The concentration of risky technology within a few modules also simplifies management problems. The most talented people can be assigned to these modules and management can thus monitor them more closely.

Low risk can be achieved in the remaining modules usually by using proven technology. In fact, companies that achieve rapid development cycles typically use proven technology in as many modules as possible. This technology may be sourced either internally or externally, but it will be proven and low in risk.

In summary, rapid product development is achieved by managing the degree and location of technical risk within a system. This risk is concentrated in a small number of modules, then checked carefully to ensure that the modules are still technically feasible. As will be seen in Chapter 12, rapid product development requires the careful management of risk.

CONTINGENCY PLANS

The modular approach discussed, and the techniques covered for controlling technical risk will normally work. But what if they don't? Does every failure have to be fatal?

Certainly not—good system design has contingency plans. Once the high-risk modules have been defined there must be some alternative in mind in case they cannot be completed on schedule. Such backup plans are normally made for high-risk subsystems. Most often they entail preserving performance, using a less cost-effective technology. Sometimes they involve backing off from a performance parameter, although this is normally less desirable because of interactions with other modules.

These decisions can be made in a businesslike manner by using the modeling techniques discussed in Chapter 2. This is done simply by evaluating the economic outcome of delivering a product late by sticking with a current plan, compared to delivering one with less performance or higher cost by using a backup plan. In most cases this analysis, which can be done in minutes, will point to staying on schedule by using the less cost-effective technology. The economics usually tend to work out this way because high cost in an individual subsystem is a rather small portion of overall costs, whereas performance shortfalls tend to have great impact on unit sales and product pricing.

RELATED ISSUES

The techniques just described generally work for most design situations. However, other issues may be encountered in the application process.

For one, the quality of the product specification can have a big influence on the success of the system design effort. A common fault in many specifications is a tendency to specify technology rather than functionality. Doing so unnecessarily constrains the system designer.

A second common problem is "creeping elegance," a design that slowly acquires more features and usually misses its cost and schedule targets. Some flexibility is necessary in every specification, as discussed in Chapter 5. However, it must be controlled because feature growth can sometimes overrun a system's margins and trigger a massive redesign. This is likely to have a devastating effect on the schedule.

One clever approach to controlling creeping elegance is to set rigid limits on a particular system parameter that will constrain addition of new features. For example, one company designing a laser printer told its engineers to keep all the circuitry on one printed circuit board. Every time the designers wanted to add a feature they ran into this constraint. It kept the feature set on this innovative product limited to the ones that were really needed. Other parameters that could play the same role are power consumption, system weight, and enclosure size.

It may be tempting to use an overall cap on product cost in much the same way. Although this is conceptually appealing, it is less successful in practice. Product cost numbers often turn out to be too squishy because of various assumptions made about quantity discounts on purchased parts, uncertain estimates of the manufactured cost of fabricated and machined components, and assumptions about overhead and labor rates. In contrast, the weight of a product, the size of its printed circuit board, and the size of its power supply are far less fluid numbers.

Architecture can also have a powerful influence on the speed of transition into the manufacturing stage. The modular approach allows well-defined discrete chunks of the product to be released to manufacturing. Normally this will allow a faster handoff between engineering and manufacturing. Subsystems that are likely to have the longest start-up cycles in manufacturing should thus be given first priority, so that the transition to manufacturing is not limited by a single slow subsystem. As covered in Chapter 13, the staggered releases of discrete chunks also smooths the glut of paperwork accompanying a new design as it moves into manufacturing.

MANAGEMENT'S ROLE

By now it should be clear that system design has an extraordinary impact on the development schedule of a product. If done improperly, it may effectively prevent any possibility of an early product introduction. It is therefore a high-leverage area for management attention.

In our experience, system design rarely gets this attention. More typically, system design and product architecture decisions are left to the specialists. These specialists often do a fine job on technical issues, but are usually unaware of the managerial issues associated with design decisions, and often lack information from management about development priorities.

Management seems to ignore this area because it is too technical, is difficult to manage, and appears to require little attention. They fall into thinking of it as an unimportant phase of the development process simply because it is done by only one or two people during a relatively small portion of the overall development cycle. The normal cost-oriented pattern is to concentrate on the program two weeks before product introduction when there are one hundred people working feverishly to stay on schedule.

The time-based manager will understand the key leverage points and focus on the fuzzy front end where he or she can really have an influence. Chapter 14 discusses more about management's role in achieving rapid product development.

SUGGESTED READING

Yourdon, Edward and Larry Constantine. 1979. *Structured Design: Fundamentals of a Discipline of Computer Program and Systems Design*. Englewood Cliffs, New Jersey: Prentice-Hall. Although oriented toward computer software, some this book's general principles can be applied to the design of other products.

CHAPTER 7

Staffing and Motivating the Team

A group that develops a new product is often called the development team. This term probably grew out of a recognition that a more cohesive group than usual is needed to get a new product to market. As the product development pace quickens, even more is demanded of the development group. Assembling this high-performance team is the subject of this chapter and the next. Here we focus on how individuals are best chosen and motivated. The next chapter describes how individuals can be organized into a group so as to facilitate communication and rapid decision making.

Teams can take many forms from one company to the next, so the staffing issues of this chapter and the structural issues in the next are closely related. Let us start by describing the characteristics of an ideal team organization and use that as a framework to discuss staffing issues. Then the next chapter will consider organizational alternatives and demonstrate why the suggested team structure will usually work best.

In our experience, the more of the following criteria that a team satisfies, the faster it will develop products:

There are ten or fewer members per team.
Members volunteer to serve on the team.
Members serve on the team from the time of
product concept until the product is in production.

Members are assigned to the team full time.
Members report solely to the team leader.
The key functions, including at least marketing, engineering, and manufacturing, are on the team.
Members are located within conversational distance of each other.

In reality, it is rare to have a team meet all these criteria. The expected markets, technology, manufacturing methods, resources available, and scope of the development effort for a new product all have to be taken into consideration in constructing the actual development team for it. For example, some products just do not have enough development work in them to justify assigning several full-time members. A company with such products will have to make some compromises in the number of people on a team or the degree of dedication possible for team members, as discussed in Chapter 8. By contrast, some projects require far more effort than even ten people can provide, a limitation that can be overcome to some extent by using the architectural strategies suggested in Chapter 6. Other characteristics of the ideal team, such as having all members report directly to the team leader, have an important bearing on team performance, but they can be difficult to implement because of company tradition or politics. Any firm that wants to accelerate its development cycle should nevertheless consider the consequences of deviating from the model.

This model, illustrated in Figure 7-1 for a typical eight-person team, is used here to consider the staffing issues associated with a fast product development team.

THE TEAM LEADER

Of all the decisions management makes in managing accelerated new product development, none is more crucial to success than the choice of a team leader. A strong leader will be able to overcome many other shortcomings and imperfect management decisions, but a mediocre one will be stymied even by small obstacles. It is important to pick a leader carefully and announce the choice publicly. Everyone should know exactly who is responsible for successful completion of the project. In our consulting we have frequently been surprised to find that people are often not sure just who has project responsibility. Management will typically offer a variety of reasons for not designating a leader. Sometimes they may say that they want everyone to feel respon-

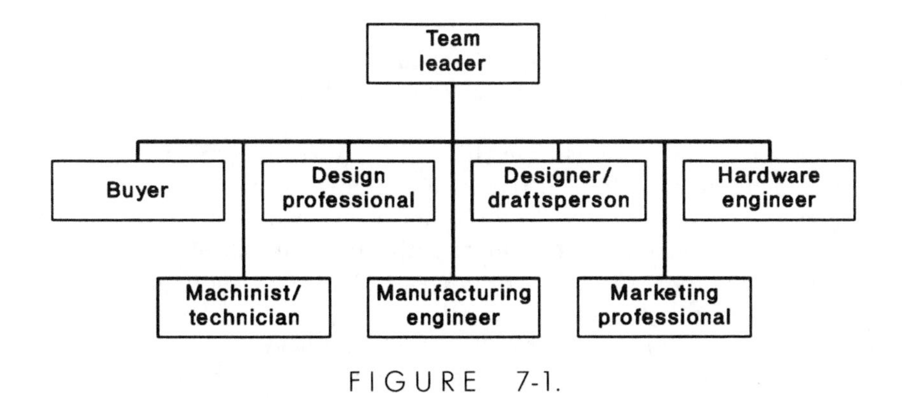

FIGURE 7-1.

A typical development team's makeup; the types of specialties required will depend on the nature of the product. The design professional can be either a mechanical, electrical, or software engineer, or an industrial designer, depending on the nature of the product.

sible for the project. They reason also that different people should lead at different times. Occasionally they are waiting for a natural leader to surface. Although these concepts are alluring, we have yet to find a case where they have worked.

For example, leaving the team leader unnamed for a natural leader to surface has two flaws. First, it wastes valuable time while team dynamics play themselves out (though this time may be in the fuzzy front end, where it is less visible; see Chapter 3). Second, it robs the eventual leader of authority by leaving his or her leadership ambiguous. In one U.S. company where this was done, we asked all the members of several teams to name the person who was leading their project. A wide variety of responses included two vice-presidents of the $500 million operation. In one project that ran into many last-minute manufacturing problems, the largest number of votes went to the manager of the manufacturing plant, which was in the Far East!

Team Roles

Many roles must be covered to make a team successful. Some must be played by team members, with others by the leader himself and some by outsiders. The most obvious roles involve ones in problem solving for specified technical areas. Beyond these relatively straightforward roles there are a few crucial but less apparent ones that must also be covered. The choice of team leader should be made with a clear plan in mind for covering these crucial roles. They need not—and in some cases cannot—be covered by the team leader, but to maximize the chances of success coverage should be provided in some form.

The first role that needs to be covered is as a product champion or sponsor. This person is typically an individual external to the team who has significant clout in the organization. The person's functional affiliation is of only secondary importance, because his or her principal function is to make sure that the organization external to the team responds promptly to the needs of the team. This champion of the cause must genuinely believe in the product, so this role cannot be an assigned one. On the other hand, the wise team leader or top executive will try to arrange for this role to be covered. In some sense this champion acts as a godfather to the project, intervening to help the team over seemingly insurmountable obstacles. Because having clout will be so important to the success of these interventions, the typical candidate should be an individual at a high level who rarely can spend a great deal of time with the team, although this champion must be accessible to the team leader when critical issues arise.

The second role is that of gatekeeper. This person serves as a window to technological news and sources external to the team and makes a point of supplying the team with the technological tidbits needed to overcome problems as they arise in the design. Although the best situation is to have each team member acting in this role in his or her own field, some may not be inclined or able to keep abreast of external developments. This activity is especially difficult to sustain in the heat of a high-pressure project where team members rarely have the luxury of gatekeeping, so that some external source may be needed. A perceptive team leader or sponsor can keep the team linked to appropriate gatekeepers.

Last is the role of technical experts. These relatively narrowly focused experts often do not fit well into a team environment and may be too specialized for the team's project to keep them occupied. Their contributions may be essential, however, for providing the key bits of innovation that can build a competitive advantage. Chapter 8 covers these individuals and their relationship to the team as outsiders.

Once coverage of the key team roles has been planned, we can focus on the special skills needed by the leader.

Team Leader Skills

Team leaders for fast development projects need skills in several quite different areas. It is most desirable to find someone who is both technically astounding and a charismatic leader. Because either of these skills is rare and, if present, will have one typically reinforced at the expense of the other, it seems unusual to find someone who has both of these key characteristics. The tendency of North American companies to avoid transferring product development people between departments also tends to keep these people's outlook relatively narrow.

A team leader is more than an excellent engineer or fine administrator. In seeking out this person we are looking for a future general manager, an individual who views the development project as a business endeavor, not simply as a technical or marketing problem. Failing to appoint such an individual implicitly makes the team rely upon a general manager somewhere higher in the organization to be the actual team leader. When this happens, decision making becomes diffuse and unclear, and the project will invariably slow up. The chosen leader must have a broad enough base of skills that management will feel comfortable having him or her alone to lead the whole project, not just a functional portion of it.

The team leader must understand the workings of the company and the personalities of its key managers well enough to be able to obtain resources and get things done. However, it is also valuable for the leader to have experience from outside the company and thus be open to alternative approaches to the inevitable problems that will arise.

Four skill areas are needed in a team leader, as illustrated in Figure 7-2. The first is people leadership skills. The development team is going to be tackling a tough job with limited time available. (If this is not the case, this is not a fast development project!) Everyone is going to have to perform at 120 percent. People are going to have to work together and operate as a team. There will inevitably be tough spots, where morale will wane. Technical, organizational, or resource crises will arise that the leader will have to detect early and make corrections for quickly, because the tight schedule will not allow for major redirection or delays.

The leader will have to be skilled at getting the best out of individuals and encouraging them to work together as a team. The ability to spot people problems and correct them early on is therefore essential. The leader will also have to establish rapport with senior management, perhaps through the sponsor, so that the project continues to have the resources it needs.

The second skill needed is vision. The leader must have a clear vision of what the product will do for the company and what determines its success. This understanding must be based on customer requirements and often also on knowing competitive products. The leader must constantly view the project in the context of its overall goals and keep the team focused on its business objectives. The leader's

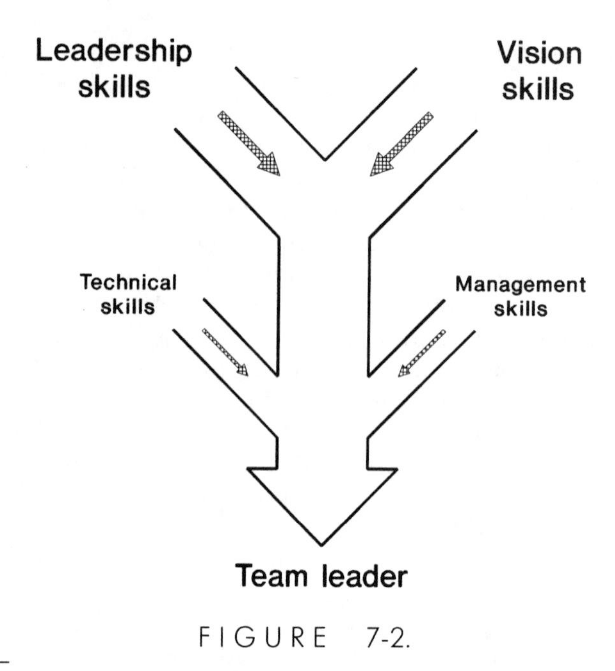

FIGURE 7-2.

A team leader needs four basic types of skills, with leadership and vision skills being paramount.

guidance is essential in a complex, constantly changing environment. The leader must resist the temptation to be attracted to the details of a project and lose sight of its objective.

Tom West, the team leader in *The Soul of a New Machine*, provides an excellent example of the use of vision skills. It was not clear to the younger members of the team what West did, but without his constant reassessments, his backstage arrangements to keep the Eagle project moving forward, and the directives he passed down to the team at critical points, the project would surely have foundered and possibly died.

There are two other skill areas required in a team leader. Although they are important, they are secondary to leadership and vision skills. One of these is technical skills. The leader must know enough about the technologies that go into the product to understand the technical issues involved, be able to identify critical obstacles, and make accurate decisions on the technology. It is appealing to have a leader be expert in the technology being used, but this is not only unlikely but may sometimes even be dangerous. It can detract from the leader's role by drawing him or her deeply into technical problem solving. Furthermore, because most of today's products involve a broad variety of technologies, it is usually impossible to keep up equally with all of them. The team leader may be trained in digital electronics but at some point have to make key decisions on, say, software or polymers.

The fourth and last skill area is that of detail management skills, often called project management skills. Someone has to manage the schedule, with all of its tasks and due dates. If the tasks are not completed on time, adjustments may be needed and resources have to be reallocated. Reports, tracking charts, checklists, and sometimes project-scheduling software are the tools of detail management. (See Chapter 10 for further explanation.) The typical new product has many parts with perhaps several variations, each of which requires different activities to be performed, especially as it moves from engineering into manufacturing. As the schedule tightens, each of these activities becomes more critical, because there is less time available to recover if anything—even something simple—fails to get done.

An excellent example of this problem was related to us recently. A sophisticated industrial product costing about $1,000 had been delayed for four months by lack of a washer! This was not a garden-variety washer but one made of an exotic material with special finish, hardness, and quality requirements. Purchasing was supposed to identify the special requirements with the supplier, and the quality-control function should have inspected the first batch of washers before they went into inventory. But even though there was no excuse for it, these safeguards failed, and ordinary washers were produced and went into inventory, awaiting the completion of the major parts. When the parts were then picked for assembly, the shortcoming was discovered, but by then it was too late to recover without delaying the whole project.

Unfortunately, this instance is not an isolated incident. When we related this incident to an executive at another company a couple of weeks later, he was not surprised—and revealed how they had just had the same experience with a screw.

Although the team leader was not the one to make these mistakes, he or she is ultimately responsible for the schedule and everything that can delay it.

Clearly, managing a multitude of details well is critical to success, but many of these details can—and should—be delegated. However, the team leader cannot completely avoid handling details by assigning them to other team members. By worrying about details, the leader will get everyone else worrying about them too. The team leader must therefore both keep in mind the big picture and be willing to dive into the details.

All too often, the team leader becomes a leader in name only and functions only as somebody to "hang" for the details of the project. This type of "leader" is never empowered actually to lead; instead, a higher-level manager retains the prerogatives of leadership and vision. These managers do not spend 100 percent of their time with the team, however, so decisions get made slowly and the team is bound to be slowed down.

In summary, concentrate on providing a team leader with vision and leadership skills. In a sophisticated, mixed technology product—which is what most products are, these days—adequate technical skills cannot be provided by one individual anyway. Detail-management skills can be learned or delegated to a large extent if necessary.

Should the Team Leader Come from Engineering?

This question, with all its political overtones, arises in many companies. Many, including respected professors in business schools, say the leader of a technical product development project must come from engineering. Others argue that only a marketing- or business-oriented person will have the breadth of vision needed to keep the project headed in the right direction for the company. Another group suggests persuasively that if the product is going to get into manufacturing quickly and at low cost, the project leader must come from there to provide an orientation toward manufacturability.

Each viewpoint is valid, but they all miss the point. What they are saying is like asking, "Should the quarterback come from Chicago?" The quarterback must come from wherever you can find the best quarterback. Team leader skills are scarce in any company. Consequently, the leader should be chosen on the basis of his or her skills, not on functional background. It will be difficult enough to find a person with adequate skills in any department, so do not constrain the search to a single department.

TEAM MEMBERS

The team leader's first job is to staff the team with the right people. Later in this chapter and again in Chapter 15 we describe how people get on the team through the techniques of recruiting volunteers. But first let us understand what kind of people are needed to build an effective team.

Rapid new product development involves much more than just engineering. The team must include members from, at the very least, also marketing and manufacturing. This statement may prove an uncomfortable one to many managers who would rather avoid potential problems of mixing engineers and nonengineers. It is relatively easy to assemble a team of engineers, who all come from the same department and tend to think and operate similarly, but getting a marketing person to join the team and participate on an ongoing, full-time basis is far more difficult. It is nevertheless essential for the fast development of a marketable product.

As explained by Robert Kelley in the reading suggested at the end of this chapter, the qualities desired of effective team members are remarkably similar to the ones that make team leaders effective. An effective follower, he stresses, is active in providing independent, critical thinking. "Self-confident followers see colleagues as allies and leaders as equals."

Team Size and Degree of Commitment

An ideal team would be staffed with ten or fewer volunteers, who should all work for the team full time. Projects big enough to require more people should use cross-functional subteams of this size. These characteristics will come under more discussion, especially in the next chapter.

Placing an upper limit on team size comes from the increasing amount of time that must be spent communicating in large teams. Doubling a team's size more than quadruples the communication burden. The communication limit can also be visualized in terms of the number of people who can sit around a conference table and get something done, or the number of desks that can fit into one room and still allow people to talk to each other. When the group gets larger than about a dozen, which is probably stretching it some, team members are likely to spend most of their time in meetings and little on developing the product. Large groups have more complex organizational structures, and it is not unusual for people to get left out, or at least feel left out, of some of the critical discussions.

Full-time participation is the key to small team size. It is crucial to building a level of commitment that will get the development job done quickly. Part-time, bit-part players lack the concentration and commitment needed to move a project along at maximum speed. Other priorities intervene, and their energy gets diverted.

A warning to managers is in order here. It will not be easy to keep some people on the team full time if the organization has not been used to having full-time commitments. Full-time assignments do reduce management's flexibility for fire fighting. Moreover, they put a spotlight on accountability that may be uncomfortable to some people. They tend to constrain an individual's freedom.

Consider an example of a software engineer on a team we helped set up. He was very talented and highly specialized in developing computer code for the machines his company produced. After a few months of working full time on the team for just one machine, he told us he disliked the arrangement, because he was used to working on several machines simultaneously. When he ran into a tough spot on one, he would put it aside and work on another for a while. This specialist, far removed from corporate planning, was unconsciously dictating the development schedules of several new products according to the tough spots that happened to arise in writing their code. He thus disliked the full-time team concept because it forced him to confront a problem and deal with it until he had licked it.

It takes conscious effort on the part of management to make sure a team stays together. Technical specialists like this one are in great demand within a company that is dependent on their expertise. It therefore becomes difficult to keep them on a team. As problems arise with existing products in the field or other new products, these talented specialists can quite naturally be drawn away from the team.

It is through the way it responds to issues such as these that management makes clear its commitment to fast product development. It will not seem "natural" to place a specialist who is in high demand exclusively onto one development team, and he or she may not stay there without some management attention. Paying attention, or failing to, to such issues also sends powerful messages to the troops about the seriousness of the organization's commitment to development speed.

Generalists versus Specialists

Anyone developing a highly technical product will need technical specialists on the team. This is especially true when introducing new technology or pushing hard on the state of the art in it.

Having a technical specialist on a team should be thought of as the exception, however. The operating style in far too many companies is to divide the development job into small, specialized pieces and farm these out all over the company. Then the schedule gets into the hands of many line managers, each having only a limited stake in the success of the project. The work then sits in a queue in accounting, purchasing, the tool room, and dozens of other places, waiting for that function's "efficient" specialized attention. In reality, such specialists become quite inefficient, because of the queue time they generate. Fig-

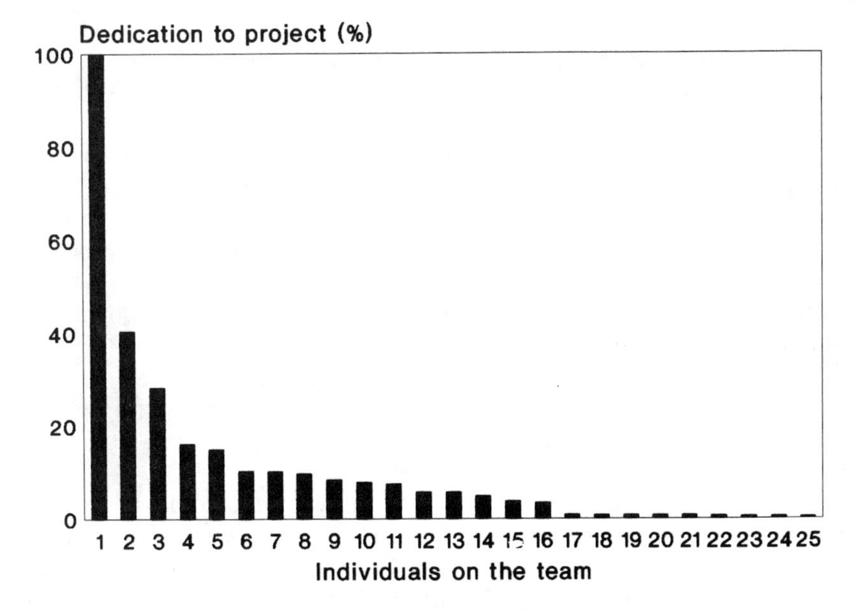

FIGURE 7-3.

In functionally organized companies teams tend to be staffed with many specialists, most having only low levels of dedication to the project. This illustration is the result of actual project data.

ure 7-3 shows the team dedication breakdown for a relatively straightforward product with about thirty parts. Of the twenty-five people on the team, only one was working on the project full time. Eighty percent of the people on the "team" were specialists with a 10 percent or lower level of dedication.

This kind of fragmentation, which has been the operating mode in North American industry for years, is deeply ingrained. Only recently have people in manufacturing operations begun to recognize that this kind of specialized efficiency is not really effective in a broader context. The result has been to shift factory operations to group technology, work cells, focused factories, and the like to reduce lead time and handling steps that do not add value to the product. The reorganization that is taking place today on the factory floor needs also to be made in product development operations.

For the product development team this means that generalists should be assigned to the team whenever possible. Doing so helps the team work more effectively in two ways. Most important, it keeps the work within control of the team members responsible for maintaining the schedule. They are the ones with a clear interest in making sure the work is done on time. The second advantage is that generalists give the team much more flexibility in keeping its members productively occupied. It is easier to justify assigning people to a team full time when there is an ongoing full-time need for their services.

Developing people into generalists with a range of skills takes more time than just rearranging a factory floor to create a work cell. But there are some things that can be done immediately to provide some generalists on a development team. For example, think beyond the normal professional, salaried person who might normally be put on the team. A person who can run a few machine tools (although not expertly), conduct some experiments in the lab, and complete a few drawings on a CAD system in a pinch can be invaluable in getting a variety of things done. A machinist or lab technician might jump at the chance to broaden his or her duties and have a key part in a new product's birth.

Being a part of a concentrated, fast-moving development team requires a certain roll-up-the-sleeves, can-do attitude, which should be one of the criteria for choosing team members. On one of our consulting projects we heard many complaints from engineers that certain people in marketing were aloof and really did not care how the product worked. So in a questionnaire we asked team members if they thought the marketing representative on their team could disassemble and reassemble their relatively simple consumer product, given a few tools. All the engineers could do was chuckle at this question. They could not picture their marketing member handling a screwdriver or a pair of pliers. The point is that not only should the marketing member be comfortable with a soldering iron, but the scientist on the team should also be able to demonstrate the product to a group of visiting customers or create a first draft of advertising copy if need be.

Understanding What Makes Engineers Tick

Regardless of how the fast development team is staffed, it is likely to be dominated by engineers, simply because engineering must play a large part in developing a new technical product. Consequently, it pays to consider the values and motivations often associated with engineers when staffing the team. Only limited research seems to be available comparing the values of engineers with those of other corporate employees, but the available data do suggest that engineers react differently. *Machine Design* conducted a survey (January 21, 1988 issue) asking design engineers to rank various job factors. Their findings suggest that being on a fast product development team is not the setting most engineers would prefer. For example, when asked to rank the factors they liked best about their jobs, interesting work, salary, and work independence rated highest, with teamwork in last place. When asked to choose between interesting work and prestige or reputation, 100 percent of the engineers chose the former. Three engineers chose to work on their own ideas for each one who preferred knowing what was expected, as, for example, meeting a delivery schedule. And four engineers chose diversified work for each one who preferred concentrating on a single project.

We may conclude that working on a single project and getting it completed on schedule is not especially appealing to engineers unless the technical work itself happens to be interesting. There are some things management can do, however, to position work so that it is more interesting. For related ideas, read *The Soul of a New Machine* (see Suggested Reading) or "The New New Product Development Game" (in Suggested Reading for Chapter 9).

Functional Parity

The development team is sure to include members from several different departments. It is important to keep them all on the same level of importance, because each will contribute essential ingredients to the new product. Many of the impediments of developing a new product have to do with the implicit pecking order that seems always to develop in a company. When all the functions are in the same multilevel building, this becomes most obvious. The lab, machine shop, and probably manufacturing are usually on the lowest floor. Then comes engineering above them, and finally marketing and the executive suite, often on the upper floors. By co-locating the team as we discuss in Chapter 8 the elevational distinctions disappear, but it is essential to establish, and reinforce, parity among the various functions represented on the team.

GRAY AREAS OF TEAM MEMBERSHIP

Regardless of how carefully team boundaries are delineated, there will be fringe cases where it is difficult to decide whether a person is on the team or not. Reexamine Figure 7-3—where should the line be drawn here? Chapter 8 covers many of these cases, because they are really organizational issues, but here let us discuss two items: a general rule to circumscribe the team, and the staffing-related topic of including suppliers on the development team.

We have found a rule that has proven helpful for concentrating on the team boundary. It not only helps to define clearly who is on the team but is also useful in keeping the team small. We ask the manager to suppose that the development project has turned out well and he or she wants to reward the team by sending all its members to the Bahamas for a week. Who would be sent? Having gray areas or excessively large teams will each cause the manager problems when it comes time to send the team to the Bahamas if the team has not been defined carefully.

Suppliers on the Team

Suppliers must be considered potential candidates when forming a team. A substantial and growing portion (presently 80 percent in some industries) of a product's value is purchased. Consequently, the need for early purchasing and manufacturing involvement, which was mentioned earlier but covered in detail in Chapter 13, is often in fact a need for early supplier involvement. As the current trend continues toward greater supplier participation in design as well as manufacturing, the need for involving suppliers as regular members of the team will increase.

Suppliers should be involved as early as possible. Through them it often becomes possible to discover the new technologies and sourcing options that need to be considered early on in a product's formulation.

Suppliers particularly need to be included as team members when the new product involves critical technologies in which the company is not expert. With product technologies becoming more complex and cross-functional, it becomes infeasible for even a large company to fund research in all potential product technologies. In a fast development project there simply is no time for the team to get up to speed on a new technology. The limitations and pitfalls of a new technology can be subtle, though, and must be recognized quite early in a rapid development program. The best way to gain this knowledge rapidly is to work closely with suppliers from the outset. Catch up with the supplier's state of knowledge by staying as close to him or her as possible, not by working through an arms'-length arrangement administered by purchasing.

Even if the company has the knowledge needed, it may still be advantageous to add suppliers to the team to accelerate development.

It would be difficult to think of a company that knows more about making computers than IBM. But when it decided to develop the PC on a crash schedule, it opted to source the two key parts of this computer, the central processing unit (from Intel) and the operating system (from Microsoft). The Intel 8088 microprocessor was available and Microsoft, by working closely with the IBM team, could develop the DOS operating system faster than could internal groups.

The next chapter suggests how to avoid the delay-prone situation of not having control over the resources needed to get the product to market because of these resources reporting to other parts of the organization. When suppliers are on the team this will be an even bigger problem, because suppliers are more remote. If this weakness is not overcome the supplier interface will become a major source of delay. We have observed these types of delays to be especially prevalent in developing such items as injection-molded plastics or high-performance castings or forgings. These types of parts are often large contributors to development time. To make the part requires a mold or die, which usually requires months of labor. Although clever computer-aided methods are now reducing these lead times, the computer methods must be applied with a great deal of process management if the schedule is really to be cut significantly.

It can not only take months to get the first parts from these new tools, but there can be an even more insidious effect on development times. With molded plastics it won't be known if the part will be satisfactory until the first ones come out of the mold at the end of a long mold-fabrication cycle. Because of flow patterns and cooling effects that depend on the molding process, the part's shape and strength remain in question until the first parts are molded. So we must go through the time-consuming process of building the mold and actually making some parts before being able to tell if the parts will work. If they don't work, much of the lengthy cycle must be repeated. For some innovative products that stretch materials knowledge to its limits it may be necessary to repeat this cycle several times. Not only does doing so add a lot of time to the schedule, but it also adds uncertainty, because it is impossible to determine beforehand how many times this loop will have to be repeated. (Some computer tools that model the process of mold filling and cooling help overcome these problems; see Appendix A.)

Because of their weaker linkages, suppliers tend to be more reactive than employees. In order to move quickly by overlapping their activities as explained in Chapter 9, team members must anticipate needs and ask for information. It is particularly important to develop linkages with suppliers that will enable them to become proactively involved in the project by making it possible for them to contribute their expertise early on and on an ongoing basis.

If a product's schedule is highly dependent on a particular supplier's ability to supply parts that meet demanding requirements, consider having one of that supplier's people on the development team. Under less demanding conditions it may be acceptable simply to have the supplier visit the team for a few days each month, have a supplier representative attend team meetings, or arrange for team members to visit the supplier regularly. The key is to get the supplier deeply enough involved in the project to feel some ownership in its outcome and become keenly aware of how delay will affect the product's success.

Outsiders Who Want to Be on the Team

Being on a fast product development team is a mixed blessing. On the one hand, members of the team often get special attention from management, and extra resources are made available to them. Things happen quickly, so there are special opportunities for learning. Team members may even be encouraged to break some rules, which may provide a certain amount of excitement.

Balanced against these glamorous advantages is a lot of extra work. New products do not get developed on crash schedules by working only forty-hour weeks. The team's members will probably find themselves doing some work that falls outside their normal duties.

Under such circumstances some people may want to tag along with the team because the glamor is at first more apparent to them than the unpaid overtime. The Bahamas Rule above applies here, so that management is clear about who is on the team. Simply tell such aspiring team members that a new team will be forming in the future and you would be happy to have them volunteer for it.

MANAGING TEAM STAFFING

The first staffing decision that management faces is one of when and how quickly to add people to the team. Invariably, teams are built too slowly. One reason is the phenomenon of the fuzzy front end, covered in Chapter 3. The market clock is running, but the project clock from traditional accounting still suggests that the project is not costing the company anything until people start charging time to it. Nonetheless, receding market opportunity may be costing the company much more than a few salaries.

The second reason managers may not assign people to a new project is much more concrete: there may be no one available to assign. This situation is a symptom of and consequence of conventional management practice, in which everybody is loaded up with plenty to do so that nobody will ever be idle. Under this system every project must start late, because of being starved for people. This backlog makes it difficult to make the transition from such an overload situation to a more market-responsive posture. People are always tied up when anyone wants to start a new project without delay.

The way out of this dilemma is to start with some hard-nosed decisions by product planning. On the one hand is the decision to select all the new product ideas it would be nice to have, dump them into the system, and wait for them to be completed eventually. Or the decision may be to pick a few sound new product ideas, apply ample resources to the projects from the outset, then manage them for quick completion. The latter approach is a difficult one to get used to, because the new-project list will not look ambitious. Product planning will not seem to be doing its job if it proposes few projects. But in the end just as many projects will emerge from the product development pipeline in either case. The difference is that with a shorter backlog at the front end the products to emerge will be fresher and more responsive to market needs, a topic covered further in Chapter 11.

It will not be hard to get a marketing person on the team early, because they are usually the drivers at this stage. The biggest difficulty will be in getting the interest of the manufacturing people, who will typically be deeply immersed in getting overdue products out the door and making the month's shipments. They will not be quite sure what they might be adding to a new product in its early stages. Chapter 13 discusses the value of getting manufacturing involved early on.

Recruiting Volunteers

As suggested, accelerated product development is a lot of work and requires a high level of commitment. It will undoubtedly require some evenings and weekends to be contributed to the cause before it is over. This kind of dedication does not occur except in people who really want to be on the team. It must be their choice. "Signing up" is what it is called in *The Soul of a New Machine*.

Being on a fast development team is neither easy, safe, nor predictable. The reason people would want to be involved in such an undertaking is that it offers some things not otherwise available in many corporate environments: excitement, freedom, a chance to learn new skills or break out of a departmental cocoon, or an opportunity to put one's name on a specific product and really mold its character.

The only way to obtain a required level of commitment to an accelerated development project is for participants to make a conscious decision to be fully involved in it. In essence, they must volunteer for the team—but they might not take the initiative to volunteer. In such a case the benefits of being on the team must be made apparent to each prospective member. Individuals will be attracted by different sets of membership benefits. The recruiter's job is to identify each prospect's motivations and present the virtues of team membership accordingly but honestly.

The team leader is recruited by management, but the leader should recruit other members of the team directly.

Rewarding the Team

As noted, the team will be expected to put in extra effort, so it is appropriate to consider what rewards will encourage and recognize this extra level of commitment.

The built-in reward called "pinball" in *The Soul of a New Machine* is to give team members an option to play another exciting game if they win this one. Over the long term, management needs to regulate the attractiveness of being on the fast development teams so that there will be an adequate but not overwhelming pool of volunteers for future projects.

The objective of this reward system is to encourage the team to achieve a certain goal jointly, which is an ambitious schedule goal in this case. The objective has to be kept in mind while formulating rewards and be used as a touchstone to test reward concepts. For example, it is essential that rewards be based on team, not individual, performance, because the objective is to encourage teamwork. The end point must be clear and comprehensive enough for all contingencies. It may not be adequate, for instance, to specify the end point as being shipping the first product if by then there is still no assembly or service documentation or training in place so that the organization is prepared to go beyond the first shipment. The issue is not that the team is trying to cheat but that the team is being encouraged to cut corners where it can, so management has to be specific about what is needed at the end. A well-written product specification will help here (see Chapter 5).

It may be worth considering a variable reward based on performance. For each month the team beats the established schedule, the reward increases by 20 percent, for instance. Before using this approach, however, make sure that the business's objectives support it. For example, if the goal is to get a product to market for the Christmas season, earliness will have limited value, but lateness would be costly. Having done a financial analysis to determine the value of time will make it easier to construct a commensurate reward structure.

The reward system should not only acknowledge what the team accomplishes but encourage others to do likewise. Thus a culture is built to support fast product development, which is one way management sends signals about its desires. Consequently, the reward chosen should be visible to others besides recipients. Cash rewards are thus poor choices for this reason.

The firm of Carrier Transicold gives us an excellent example of providing this visibility. They used a fast development team for their Phoenix semitrailer refrigeration unit. The team was highly successful, developing in six months a product that would normally have taken two years, with a product that won a prestigious design award in addition. Carrier wrote up the project not just in its own company newspaper but helped get it into a national engineering magazine. The parent company, United Technologies, then went further by including a full-page photo of the three team leaders in their annual report. Beyond this, United Technologies placed full-page ads with a photograph of the team leaders in national publications, including the *Wall Street Journal*. This is priceless recognition for a job well done. The company's main objective was undoubtedly to publicize its new product, but at the same time it sent powerful messages to its employees about what it valued. One message to be considered seriously in a case like this is whether the photo should include team members as well as leaders.

A reward will have more value if it has some lasting significance. This limitation is another shortcoming of lump-sum cash payments. Unless they are substantial, such as being large enough to buy a new car after taxes, they are likely to be deposited in the checking account and dribbled out to pay the orthodontist or mechanic. It is probably more effective to convert cash into a vacation reward, perhaps with corresponding time off. The vacation will not only be remembered for years but may also recognize spouses who, in a sense, also contributed to the project's success.

Implicit in the discussion so far has been the presumption that all rewards are in addition to regular pay, which suggests that service on a fast development project is beyond what is normally expected of employees. Most companies would prefer to have rapid development be simply a normal part of a job, not extraordinary service. The members of a fast development team will work harder, though, which justifies

some degree of extra recognition or compensation. Be careful, however, not to create an elite corps resented by other employees. Even if the development team is largely self-contained, it will still have to work through other employees to commercialize the product. A project can be torpedoed easily if the team is perceived as receiving unjustly favorable treatment.

We have tried to provide here some ideas and guidelines about rewards, but the best solutions are those that are custom created to fit particular circumstances. Your own may be quite different from anything suggested here. We are reminded here of a client who created such an effective motivator for his team. He wanted to make sure his team focused on the proper end point, so together we considered various rewards. The custom solution he chose was not so much a reward per se, but it did achieve its objective well. He called upon a former engineer from the organization who was exceptionally talented and greatly admired by the employees. This engineer had risen quickly through the corporate hierarchy to become a group president of the *Fortune* 500 parent company. Our client simply announced to the development team that he had invited this group president, their former colleague, to visit the plant on the scheduled completion date to watch the product operate. The date was on the executive's calendar, who was too savvy to have been buffaloed by an incomplete product. Besides, the team would gain sincere recognition by being able to demonstrate their creation to an executive fully capable of appreciating its performance.

Project Termination

Every project must one day end. For the manager, project completion is a double blessing: the product is now available to sales, the team members available to start something else. Unfortunately, the team members may not see it quite this way.

A true fast development project becomes an intense effort for dedicated members of the team. If a project has been properly managed for speed, the team's members will have had little time to think about anything else. The project will have become their life. Now, with the exception of some mop-up activities to stabilize the production process, this life will be coming to an end. The more intense the effort was, the more abrupt and unanticipated the ending will likely be for them.

A few months before a project ends, management needs to start planning follow-on opportunities for the team members. The available options should be discussed with each team member fairly early. Doing so lessens the shock at the end and allows an opportunity for each team member to catch up on deferred activities and become rejuvenated for the next assignment.

Membership on a fast development team creates a valuable set of skills and perspectives. Management has both the responsibility and an opportunity to redeploy these resources to further the objective of fast product development while also avoiding burnout in key people.

Once a company gets into a pattern of developing products faster, this transition will become more natural. It will be understood that the company does provide well for those on fast development teams. Employees will then be able to start on another project more easily, because short development cycles will have given them recent experience in starting on a new product. If the development cycle is long, people forget what to do at the beginning of a project, because they have not been through it for a long time.

These, then, are the kinds of people who need to be on the team. In the next chapter we examine how to organize them to get the job done quickly.

SUGGESTED READING

Kelley, Robert E. 1988. In praise of followers. *Harvard Business Review* 66(6): 142–48. Effective followers are those able to think for themselves and work without supervision.

Kidder, Tracy. 1981. *The Soul of a New Machine*. New York: Avon Books. This story of the development of a new computer illustrates many of the points discussed here on staffing a development team. One shortcoming, though, is its suggestion that engineers develop a product without interacting with other functions. *Must* read for anyone who has never been on the inside of a development team. Also available on audiotape from Books on Tape, PO Box 7900, Newport Beach, CA 92658.

Roberts, Edward B., and Alan R. Fusfeld. 1981. Staffing the innovative technology-based organization. *Sloan Management Review* 22(3): 19–34. Excellent discussion on the various roles to be provided for in staffing a development team.

CHAPTER 8

Organizing the Team for Action

In the previous chapter we concentrated on the importance of staffing the development team with competent, motivated people and suggested how such people could be selected. If one of these people alone could bring the next new product to market, that chapter would have been sufficient. Unfortunately, even if new products could be designed by just one technical person, they would still require the interaction of many people in the company to complete their journey to market. Once the right people are in place, the next issue is how to interconnect them so that continuity is maintained across functional interfaces. In electrical engineering this is called making wiring diagrams, and matching impedances. In mechanical engineering the current terminology is "minimizing the number of fasteners." Much of what was discussed in Chapter 6 regarding system architecture needs now to be worked out here at the human level as well as the hardware level.

Team structure is critical to development speed because it affects three key areas: commitment, communication, and decision making. These are the three areas that have everything to do with getting the product out the door quickly.

In our consulting work we have seen organization after organization where senior officers are urgently awaiting a new product. For those doing the work, however, the new product is just another task in the in basket. It will get done later, tomorrow, if there is not enough time today. The sense of urgency highly evident in the executive suite becomes greatly diffused by the time one visits the lower rungs of an organization. Sometimes the people in the lab are even working excitedly on something that has nothing at all to do with the future of the company. The whole purpose of organizing for product development is to focus this energy on one single product until it is shipped, then transfer the accrued energy to the next new product. In this way the new product becomes something more important than just one of many other tasks in an in basket.

Countless small details about a product need to be communicated among members of the team to make progress on a design. The manager's job in designing an organization is to maximize the number of decisions that can be made without extensive communication, thereby minimizing the amount of information that must be communicated, and creating communication channels that are quick and noise-free. The faster and more clearly information can be transmitted, the faster decisions can be made and the sooner a project can be completed. Faulty communication usually manifests itself as either delayed decisions or poor ones that result in unnecessary design reworking.

The main purpose of communication is to support decision making. A new product takes form from a string of decisions, often ones that are highly interdependent. If these decisions are not made by the people working on the project daily, the project must be delayed every time such a decision has to be made. Sometimes such decisions are simple, but they may still have serious ramifications. To eliminate delay every time a decision is needed, minimize the need for external communication by giving the team the resources and authority to make the vast majority of product decisions itself. New products are not slow in creation because it takes a long time to make a drawing or construct a prototype, they are slow because it takes a long time for the appointed parties to make a decision on what should go into the drawing or decide from whom to procure parts for a prototype. Appropriate organization and delegation of decisions can short circuit many of these delays.

FORMS OF ORGANIZATION

Modern managers recognize that a functional structure for companies thwarts cross-functional activities such as product development. Functional organization has many advantages that stem from its task specialization and division of labor, but it has important disadvantages for rapid product development. The management challenge is to provide a structure that will retain the strengths of functional management but compensate for its weaknesses.

Functional organization can be effective for stable activities that take place in a single functional area without needing extensive communication with other functional areas. The manufacturing of established products falls into this category. Functional organization also has great strengths for an R&D organization that is working on fundamental advances in new areas of technology, because it keeps technical specialists bunched together where they can support each other.

Nonetheless, activities such as rapid product development often need their own different organizational forms. Because product development requires cross-functional activity, it will probably need a hybrid structure, which is a completely different concept than a matrix one.

Any time we talk about organization we must recognize that a formal organization chart is just an imperfect representation of what may be a far more complex set of informal organizational relationships. The superior–subordinate relationships in an organization chart may not be the most important factor in determining how it operates on a daily basis. However, these charts can be used to represent the underlying connections in an organization and the nature of the authority delegated to those charged with getting a product out the door.

From Functions to Teams

As mentioned, organization diagrams can assume all sorts of convoluted forms that make them difficult to decipher. Further confusion has been added by the use of dotted lines to show lines of authority. The manager of quality, for example, sits in manufacturing reporting to the director of production but with a dotted-line relationship to the president. Does the dotted line influence the manager's ability to stop production by rejecting a shipment of defective components? The organization chart will not tell us, but reviewing specific instances of how the relevant parties have interacted in the past will provide clues as to who really wields the power.

To us there is one key parameter that distinguishes the effectiveness of rapid development teams: the relative authority of the team leader and functional manager over the resources and design decisions to be made about the product. A look at five examples along the spectrum from functions to teams will provide us with the insight to distinguish fast-moving forms from inertia-bound ones.

The first form, illustrated in Figure 8-1, is functional. A project is divided into its functional components, with each component assigned to its own appropriate functional managers. Coordination is handled

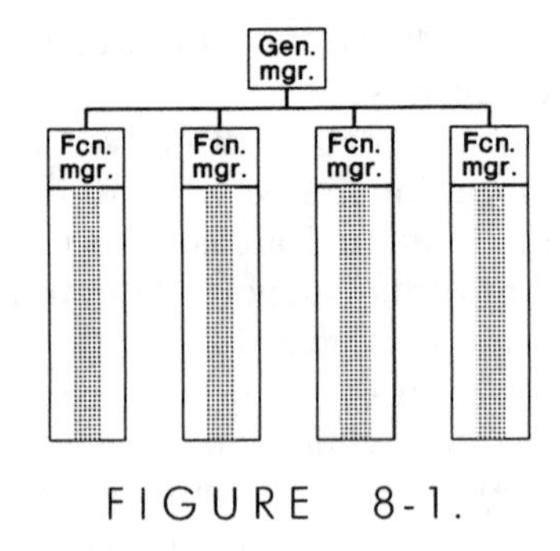

FIGURE 8-1.

A functional organization, in which authority rests with the functional managers.

either by them or upper management. All projects are basically managed by the same management group, so they all tend to get standardized treatment. Standardization can be an advantage if the organization wants to use uniform components and methods, but it can also dangerously mix urgent and routine projects to the point that it becomes impossible to rush anything.

For product development this form has generally been superseded by the other, more team-oriented approaches to be described, yet there are two cases in which it has some merit. First, where a company works on only one new product at a time, it allows managers to concentrate on the project at hand. Second, when a company is developing scores of new products at once, none big enough to justify having a dedicated team leader, the functional form ends up being used by default.

The next form of organization, and probably the most dangerous one, is the lightweight team leader form. As shown in Figure 8-2, there is a nominal team leader in this case, who is devoted to the project. The problem is that this person has no real power and just oversees plans created by others. Some companies recognize this either consciously or subconsciously when they call such a person a project coordinator.

This form has the ability to move a project along faster than the functional form can, because there is one person who feels responsible for drawing up schedules and checklists and monitoring compliance. Details are thus not so likely to be forgotten.

The lightweight team leader by definition has little power to make changes or reallocate resources, however. In this person responsibility and authority are poorly matched. The functional managers, from their experience, seniority, or political acumen, retain authority over the

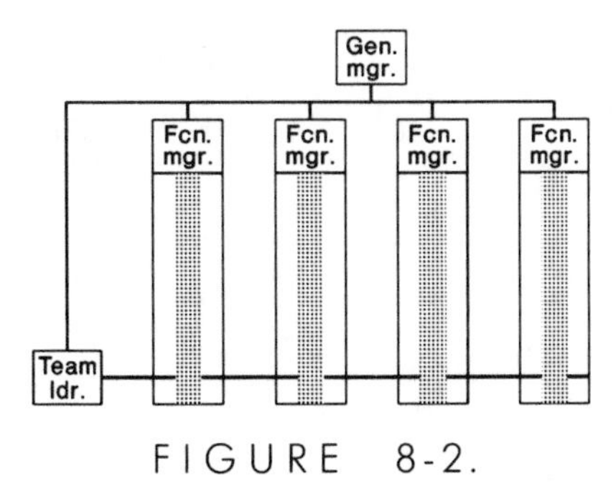

FIGURE 8-2.

The lightweight team leader form, in which team leaders exist but have little authority relative to that of functional managers.

project. Communication in this form of organization is just as circuitous and decision making as slow as in the functional form, because any real decisions must involve the functional managers as well as the team leader, who helps carry out the plans.

The danger in the lightweight team leader form is that the companies that use it usually believe that they are using a more powerful team approach. They moved from the functional form to the lightweight team leader form and now think they have a project team. What they have yet to realize is that they have taken only one step out of the three or four they need to take. Nor do they realize that the effect of what they have done has been to add one more layer to their old bureaucracy.

We call our next organizational form the balanced one because it makes an attempt to balance the power of the functional managers and the team leader. This arrangement is often called the matrix form, but that term covers a range of everything except the two pure extremes with the functional form at one end and the separate project at the other (see Fig. 8-3).

There are various ways to divide up the authority of the functional managers and team leader. One is to give the team leader control over how an individual's time is to be spent and the functional manager be responsible for the individual's professional growth and development within the company. The functional manager thus concentrates on developing human resources within a functional area. Another approach is to give the team leader control over project-related matters and let the functional manager retain control over developing functional expertise. For example, in the mechanical design function the functional manager may guide philosophies on the stress analysis methods to be used or safety factors to be employed with them, or

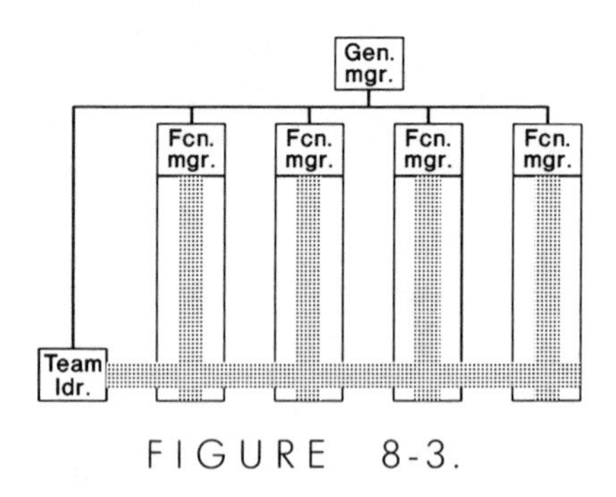

FIGURE 8-3.

In a balanced matrix the authority is shared between functional managers and the team leader, but the division of power is unclear.

might direct the use of standardized fasteners across a company's entire product line.

This type of uniformity across a business is certainly of value in enhancing product quality and serviceability, and it can help maintain a consistent image across a product line. Unfortunately, within the project team it is exceptionally difficult to establish appropriate boundaries between functional technical guidance and design freedom. Consider a case where the functional manager wants to use a fastener used by many other of the company's products, but design trade-offs on this particular product suggest using instead a special fastener that will accommodate certain automated assembly machinery that the team plans to procure. In a case like this the functional manager ends up being drawn into the team's product design decisions, which enlarges and slows down the decision-making process.

In principle, the balanced form seems to offer the best of both extremes. In practice, though, it is a difficult form to manage because it is hard to separate the areas of authority. Much time can be wasted in working out the turf issues. Matrix management has in fact hamstrung many organizations.

The next organizational form, illustrated in Figure 8-4, has a heavyweight team leader. Here the leader controls all the project-related issues, including design trade-offs, but the functional manager retains title to the people. The functional manager continues to write their reviews while they are on the project. When the project is over, these people return to their managers.

This form by definition gives the team leader the authority needed to get products developed quickly. It avoids the type of ambiguity just described over who has authority to specify fasteners, for instance, by clearly assigning all such decisions to the team leader. This form can

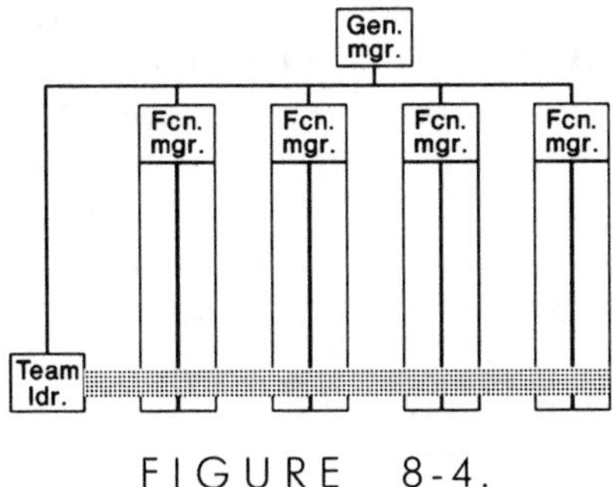

FIGURE 8-4.

The heavyweight team leader form provides the team leader with clear authority over the team members on the project.

also provide professional coherence and assure that there will be a home for team members when a project is over.

The last form of organization, the separate project, is shown in Figure 8-5. Here the project team is truly an independent unit within the organization. Relative to the heavyweight team leader approach this one can foster a more independent, entrepreneurial environment in which time to market can be a do-or-die concern. The Eagle project in *The Soul of a New Machine* was of this type.

Venture groups are an example of separate projects, but they tend to be more than just projects, because they carry their own full business and administrative responsibility. As a routine corporate mode of operation the 3M Company spins off venture groups to bring new products to market. However, venture groups are often not the fastest way to develop new products—3M is not renowned for its speed as much as for innovation—because developing a business structure takes energy away from developing products.

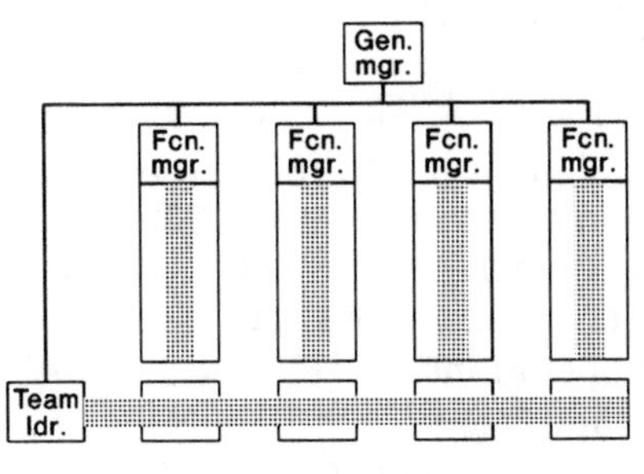

FIGURE 8-5.

The separate project form severs ties to the functional departments.

The original skunk works, established by Kelly Johnson at Lockheed, was a separate project, although it continues as an organization to this day. Skunk works have been tried by many others, with outcomes ranging from very good to very bad. There has been much variation and a consequent variety of opinion on skunk works. The quality of its leader, the way in which senior management interacts with the team, and the success the team has in operating effectively without alienating itself from the rest of the organization are all factors in the differing assessments of skunk works.

Beyond an individual project, the success of a skunk works structure in a company depends on the team's success in establishing a new, viable form for managing a project. Too often a skunk works is set up because all other approaches have already failed and the only way to complete the project is to go outside the regular system. The skunk workers become renegades who have lost respect for the company, and the company casts them off. This approach might get one product to market, but it does not build a fast-cycle capability that will add to a company's competitive strength. It is a short-sighted expediency.

The separate project should be considered an alternative development process, with its own appropriate control systems, that is established by management for use whenever circumstances suggest it. Just as some employees, children, or students need special treatment in order to blossom, some development projects also need to be put on their own special track to meet a particular market opportunity. A company needs both a Method A, which is probably some sort of a matrix, and a Method B, which may be a separate project, to be responsive with new products. Method B will not be universally better, only better in certain cases. If it is executed properly, it will be too demanding of talent, resources, and management attention to be used on every project.

Although the separate project has its price, it provides two important benefits. Most clearly, it enables a company to respond quickly with a new product when necessary. More importantly, it provides an ongoing means of challenging and rethinking the Method A approach, to see if it is as good as it can be. It is an experiment in effective new product development with the ultimate power to revitalize the company. If the separate project approach is used, management should review it periodically to make sure it is enhancing the company's capabilities and consider what lessons can be transferred from it to the company's regular operations.

Creating a Method B approach creates the potential for animosity. Method B people can be recognized as an elite group, which may encourage others to try to sabotage their efforts. The solution to this dilemma is consciously to control just how elite the group is allowed to be. Others should admire Method B and aspire to its level of proficiency. Others should want to volunteer for the next team if they are good enough. On the other hand, it should be clear that operating under Method B is not an easy life. Method B people may wear sweatshirts rather than ties or blouses and show up at noon, but they may also work until three A.M.

Finally, Method B sends definite signals, which should be thought about. For example, one of our projects once got named a fast track team. Rumor quickly got around the building then that everyone else was on the slow track, which did nothing to speed up product development.

The Balance of Authority

As we have said, the distinguishing feature among these five forms of organization is the authority of the team leader relative to the power of the functional managers. Fortunately, the team leader does not have to gain his or her power by matching the functional managers' seniority and expertise. Instead, a lot of the authority can—indeed, must—come from top management support, as discussed in Chapter 14.

The nature of the project leader's authority is the essence of project management. The project leader must have the power to make decisions and appropriate resources. Too often, project management is equated simply with the tools of the trade, scheduling, planning, and the like. One can buy software for a personal computer that will do the scheduling; that is the easy, insignificant part. Management must empower the team leader with the authority to carry out the schedule.

Just as the team leader must be a doer, not a coordinator, the team members must be the actual doers of the work. On one consulting assignment the company had impressive posters placed around the building promoting the value of teams. We were thus expecting to learn something about teamwork, but the "teams" turned out to be really committees, with their members really coordinators. The team member from purchasing, for example, was a manager who got assignments and took them back to the buyers of all the specialized commodities that had to go into the product. This "team" just added one more layer of management and further complicated already existing communication patterns among those who had to get the work done. Teams are the antithesis of committees.

Specialists and the Team

Manufacturing organizations tend to evolve into highly functionally specialized structures. For stable, ongoing production this mode can be efficient. It is also effective in complex, leading-edge research where the only way to make progress is by using highly specialized technical people. Military aircraft and supercomputers are such examples. Most product development falls between these two extremes of stability and radical departures, however. In this middle ground the use of specialists slows progress, especially if they are only partially devoted to the project.

Turn back to Figure 7-3, an actual example of a development team at a company highly optimized to serve ongoing production. Each bar in that illustration represents a person on the team, with most members clearly devoting less than 10 percent of their time to the project. Communication would be easier and more effective, decisions be made faster, and higher levels of commitment to this project as being something special would exist if this fragmented crew were to be replaced by three or four full-timers.

There will always be situations, however, where part-time specialists must be used. These conditions are of two types—fringe areas that must interface with the rest of the company, and technical specialists who really don't perform well on a team.

Fringe players are just the people we would like to eliminate from the team, but occasionally they play a critical though small role that cannot be played by another team member. Depending on the nature of the product, these people could be in finance, marketing, quality, or a similar area. (Note that many areas, such as purchasing or tooling design, can often be handled by broadening the responsibilities of regular members of the team or by adapting other systems.) The solution is to establish a well-defined connection with the team on a regular basis. The fringe player may have a desk in the team area and sit with them one day a week, or every day from 8:00 to 9:00 A.M. When the time comes, the fringe member joins the team, even if this means bringing along other work to fill the gaps. Each situation is different, so it takes some effort to find the right solution.

Support groups such as drafting, the engineering laboratory, or the model shop often fall into the fringe player category. The first option here should be to get them out of the fringe group by finding a broadly skilled individual who can justifiably be assigned to the team in a full-time capacity. After exhausting this possibility, consider the solutions just suggested for use with true fringe players.

The technical individual contributor should be a rare person with key expertise to contribute technically but who for one reason or another just does not fit into a team environment productively. The best approach here is to define the job to be done as a complete package having only minimal interfaces but definite milestones. Then have someone on the team act as a liaison on a regular basis to make sure the work meets the requirements and stays on schedule.

The fringe player and the individual contributor always represent problems. The first alternative in each case should be to complete the project without this person, or else put them on the project full time. The Bahamas Rule in the previous chapter will help in making clearer decisions about part-timers.

End-to-End Involvement

Team members should be involved from beginning to the end of a project. It is difficult to get everyone's participation at the outset because some people will be cleaning up details from the last project or be trying to disengage themselves from another, lower priority, project.

Many people will try to claim that they do not have to be involved at the beginning, because their specialty is not needed until later. For example, manufacturing engineers may say that they cannot do much until there are drawings to look at, and software engineers like to have the hardware nearly complete, to provide them with a stable environment for doing their software design. To compress development time, recognize that there are several fallacies with this staggered-start approach. One is that the whole concept of accelerated development is simply to get the momentum up as quickly as possible—and keep it there. No team can get up to speed if half the spark plugs are missing.

Another fault with the partially staffed start is that some fundamental design decisions—the most fundamental ones, in fact—are made on Day One, such as whether the product's case should be molded from plastic or fabricated from sheet metal (a manufacturing–engineering issue) and whether certain product features are to be implemented in hardware or software (a software issue). If these people are not involved at the beginning, it will deprive the project of some needed expertise, and those who will have to execute the decisions will not have had opportunity to sign up for the decision itself. These omissions will lead to delay later.

To start some of these activities early will require new behavior out of some team members, who may need help in looking at the project differently to find ways of being productive from the start. This gets into the overlapping activities approach discussed in the next chapter.

Management will also have to change some of its habits. A key tenet of management is to keep everybody busy doing something all the time. Do not assign the best people to a project full time until being sure that there is plenty to keep them occupied. There are two fallacies to this mindset. One is that the objective is not to keep people busy: it is to make them productive. It is better to have someone sitting around sipping coffee for the first few weeks of a project if this prepares them to work 50 percent more effectively later on. The other fallacy is to think of people as being so specialized that they can do only one thing. The team must be staffed with people willing to learn new skills. Let the designer figure out how to install the CAD terminal before there are drawings to make. Let the manufacturing engineer learn how to do some capital equipment planning for the machinery that will have to be procured later. Both of these specialists may be moving furniture for a couple of days.

As a project gets into its later phases, there will be a great temptation to phase people out. Then new people will take over after a handoff. But handoffs destroy momentum because the enthusiasm for the product that was developed by the original team cannot be transferred to successors.

There are generally two reasons why the original team leaves a project as it starts to move into manufacturing—one related to management and one to the members themselves. Management of course wants to free up the development people to start a new project. When making these decisions managers seldom recognize, though, how many details are still left for a new crew to clean up and how long it will take them to become familiar enough with the product to do so effectively.

The design and development people will themselves often try to bow out as the product leaves engineering. In North America, manufacturing tends to have lower status than engineering, so design engineers would just as soon avoid the factory floor. And in many companies the manufacturing startup process is bewildering. There are complicated computerized production control systems into which each part has to be entered, and few people, even within manufacturing, understand how these processes work. In manufacturing one often has to work with unionized workers over whom one may have only limited influence. It is thus more comfortable to stay in the design room and start on a clean sheet of paper.

Xerox has recognized that handoffs can cause delays in product development and has therefore structured its development teams to minimize them. Its development teams are headed by a chief engineer with cross-functional experience who may in fact come from any function. Reporting to the chief engineer are managers of all the functional groups needed to develop the product. To integrate product design and manufacturing, one of these direct-reporting people is the manager of a manufacturing resource team (MRT). This team is responsible for manufacturing engineering, manufacturing quality, and materials and spares management related to the product. By focusing on issues like design for assembly and just-in-time processes, the MRT ensures that the design is suitable for efficient, ongoing manufacturing. As the product moves into manufacturing the MRT moves with it to provide a smooth transition. The MRT manager then becomes the ongoing production manager for that product. There is thus no manufacturing handoff at Xerox! The shared values and broad accountability of the development team tend to keep its members focused on making smooth transitions through all the phases of development from the initial product concept to delivery to the end user.

The Team's Executive

There are two difficulties in setting up a dedicated joint development team, or separate project, in a highly functional organization. One is in interfacing with the inevitable fringe players, as discussed above. The other is in picking an appropriate person to whom the team will report. Because the team is intentionally multifunctional, to overcome department barriers, it is inappropriate for the team to report to the head of engineering, manufacturing, or marketing, especially if there is the normal amount of polarization between these functions.

The most obvious solution is to have the team leader report to the general manager. This manager is usually pretty busy, but this new arrangement can be a welcome change of pace for him or her. In one project that adopted this approach the team leader was not particularly outgoing and was a bit uncomfortable creeping up to the thickly carpeted executive suite in his jeans and sweatshirt to discuss the project with his boss, the general manager. Fortunately, this did not have to happen too often because the general manager usually visited the team, which was easy for him to do, with the whole team located in one room.

In theory, it is possible to find an executive from another function, such as finance or human resources, to shepherd the team, but in practice such people usually have little interest in or knowledge of new product development and may also not have the organizational clout of a general manager.

CO-LOCATION

Once the team has become organizationally distinct, it is important to reinforce this status by also making it geographically distinct. Put all the team members together in one location. Physically moving team members to a new location helps them break away from old patterns and establish new ones focused on development speed.

A lot has been said about the "walls" that exist between departments in functionally organized companies. "Tossing it over the wall" is a phrase everyone understands about the development process. The obvious solution for our co-located team is not to have any walls. Marketing, engineering, design, manufacturing, and other functions work together in their own wall-less area.

Actual walls are helpful surrounding the team area, to provide separation and shield the team from intruders. One team we worked with not only put a wall around their area but even put a combination lock on the door to keep all—including the boss—out. After a while the door stayed unlocked and was even left open, but in the beginning it served to set them apart from the rest of the organization and shield them from distractions. People visited the team only when they had business to conduct and were escorted out when it was over.

The Communication Factor

Co-location is initially helpful in punctuating the shift to a new mode of operation, but its major value is of a longer-term nature. When the team does get to work, having the team members sit side by side greatly enhances their communication, which in turn greatly speeds up product development.

Product development involves a great deal of seemingly trivial communication and decision making. There are, for instance, countless possibilities and ramifications for a new design. The direction in which it ultimately goes, or should go, will depend on the inclinations and background knowledge of those involved at the time. If technical people are the ones making the key decisions, the result is likely to be technically elegant. If marketing is deciding, the design will presumably be market driven. And if manufacturing decides, the product will probably be cheaper to make.

Industrial managers know that the key product development decisions and trade-offs must involve several functions, which then do eventually get involved, even in a highly functional organization. But they get involved in a slow, circuitous way, sequentially and plodding along every step of the way. In manufacturing jargon, there is a lot of "scrap and rework" in the design process before a quality product results. We must accelerate the communication process to speed up product development.

Again, proximity greatly enhances communication. The big, obvious issues can always be communicated formally, but for each of these there are many, less obvious, issues that are fragile. If the communica-

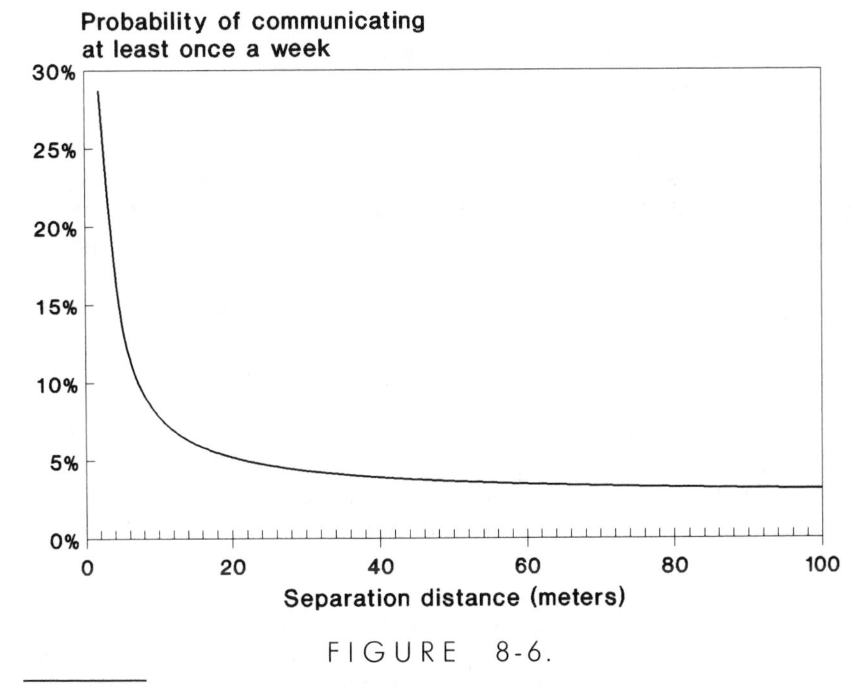

FIGURE 8-6.

Technical communication is much more likely to occur between team members if they are located close together. *From Allen,* Managing the Flow of Technology, *Figure 8.3, p. 239, The MIT Press, 1977.*

tion links are at all restrictive, these issues will fail to be communicated. Professor Thomas Allen at MIT has done some interesting research regarding the effect of distance on technical communication. By measuring the frequency of communication of 512 individuals in seven organizations over six months, he developed an analytical relationship between technical communication and distance. That relationship is shown in Figure 8-6. The knee of the curve is at a surprisingly short distance (ten meters, or about thirty feet). Beyond about thirty meters (100 feet) the curve is so flat that further separation becomes immaterial. It is clear that team members must be located close together, definitely closer than ten meters, to communicate effectively.

Professor Allen provides no explanation as to why communication drops off so rapidly with distance. In our consulting, though, we have observed how communication occurs in the product development process and can suggest an explanation. A great deal of communication occurs in co-located teams simply from the ease of hearing another person's conversation directly, even when it is not initially intended for you. When all of a team's members are located close enough to each other to be exposed to what their colleagues are saying, much of their colleagues' information becomes theirs automatically. Team members then begin to assimilate other points of view and work them into their own understanding of the product. When discrepancies occur, they get discussed immediately and a solution can be found earlier, with less time being wasted by working on the wrong concept. Much of the advantage of a closely knit development team is this sharing of partial information so that energy gets focused on the most critical areas. (The concept of partial information is covered in detail in the next chapter.)

We have also seen cases where an initially strong team lost this ability for fast communication because key members started to drift away. When these members, particularly if they are in marketing or manufacturing, stray from the group, its decisions no longer reflect the marketing or manufacturing considerations that had initially been a

part of the design. Because a co-located team can indeed move quickly, if its key participants are not within listening distance the team can quickly start moving in new directions that will unconsciously exclude those participants.

As noted, team members can drift away. This is one factor that management needs to monitor, particularly if there are key part-time members. The problem is that the organizational strings tying members to their old departments never completely dissolve and subtly keep

pulling them back "home." We recall one team that chose to build its prototype machine in the team area and install a small machine shop there to do that. The marketing member, in his three-piece suit, ended up sitting within thirty feet of a Bridgeport milling machine, which is all too audible. The carpeted, partitioned area back in marketing seemed quite pleasant in comparison. It took some work to keep that member involved with the team. (Incidentally, the team chose next time not to co-locate themselves with their private machine shop.)

Although this marketing example may be the most graphic one, this problem affects other functions as well. Purchasing, manufacturing engineering, and production control in particular, tend to be highly specialized areas that have developed detailed systems for doing their jobs. Modern computer technology allows us to move some of these systems to workstations within the team area, but some elements of the system, such as the information stored in the heads of colleagues, cannot be moved. It is not always natural for a team to stay together of its own will.

Professor Allen's communication–distance curve (Fig. 8-6) suggests that beyond thirty feet of separation co-location provides little benefit. We have observed, however, that just getting all the team members under one roof can help greatly. Here is just one example to illustrate the magnitude of a significant communication improvement. General Motors went to considerable expense to build a new engineering center for their Flint Automotive Division, co-locating four former engineering groups that had been separated by fifty miles. Formerly, if separate design groups needed to make a trade-off decision, they would schedule a meeting about three weeks in the future (the travel time between locations required half-day blocks for meetings, so they had to be scheduled far in advance). Typically, a meeting would be cancelled and rescheduled once, so it took six weeks to resolve just one issue between subsystems.

After all the engineering groups were under one roof, Gary White, the leader of the suspension design team, encountered a problem of insufficient clearance between the car's wheel and the inside of the fender (the "wheelhouse"). On his way to lunch he stopped by to see Dennis Murray, the leader of the body structure team, in his office. In five minutes they had the problem resolved and went to lunch. The time savings was 99.99 percent.

Separate Facilities

There are strong reasons for the development team to be located together but be somewhat isolated from the rest of the organization. The appropriate degree of isolation depends on two factors. On the one hand, the team would like to be close to the information, facilities, and experience available in the main body of the organization so that they can make use of those resources. On the other hand, part of the advantage of having a separate team is to overcome some of the bureaucracy, delay, and control systems in the main organization. At one mature company an employee told us that delay was built into the very walls; the only way to get anything done quickly was to get out of the building.

Having a separate site does remove many of the distractions and onlookers that can plague a new concept at home. Some isolation can engender a sense of independence coupled with the feeling that there is nobody else around to lean on if things don't work out. The team can then feel more responsible for its own destiny and be in a stronger position to control it.

When IBM decided to develop their PC personal computer quickly in 1980, they knew that the project could get bogged down by the overabundance of computer knowledge that existed within the organization. To move quickly they saw they would have to work closely with suppliers and make use of as many non-IBM parts as possible. This was not the IBM way at the time, and and it became apparent that design philosophies could be debated endlessly while competitors slipped into control of the personal computer market. So IBM established the team in a warehouse in Boca Raton, Florida, far from corporate headquarters and other corporate development facilities.

Usually the development team wants to be located close to the main facility so that they can run back to get information from colleagues and borrow tools and materials. A few years ago technology put a limit on a team's ability to move far, because mainframe-based CAD systems then placed limits on data-transmission distance. However, with the improving capabilities of PC-based CAD systems this is no longer much of a restriction.

In the end, it is the team leader with the team's concurrence who must decide where the team needs to be located. The goal in forming the team should be to provide it with the resources it needs and remove any external reasons the team can identify as being possible sources of delay. If the team thinks that being in its regular building will be a hindrance, it probably will be.

The communication issues considered in this chapter take on vital importance as we move onto the next one to look for opportunities to execute activities in parallel, to save time. As we will see, having parallel or overlapping activities requires participants to work with only partial information, because the tasks from which they must draw their information will not be completed yet. The only way this kind of fluid situation can be productive is to have open communication channels so that information stays up to date. Consequently, communication issues receive a great deal of emphasis here because in addition to their direct value they are critical to the successful overlapping of activities.

SUGGESTED READING

Allen, Thomas J. 1977. *Managing the Flow of Technology*. Cambridge, Mass.: MIT Press. Interesting findings on the effects of communication and structure on R&D effectiveness. Chapter 8 in particular discusses the relationship between physical separation and the frequency of communication between technical workers.

Allen, Thomas J. 1986. Organizational structure, information technology and R&D productivity. *IEEE Transactions on Engineering Management* 33(4): 212–17. Allen argues that the decision between using functional or project structures depends on how rapidly the company's technologies are changing. If they are evolving rapidly and the company's technologists are sequestered on a development project for too long, the company could lose its technical expertise.

Hayes, Robert H., Steven C. Wheelwright, and Kim B. Clark. 1988. *Dynamic Manufacturing: Creating the Learning Organization*. New York: The Free Press. Chapter 11 of this book focuses on the management of product and process development and provides a particularly good discussion of the pros and cons of various forms of organization. The authors follow the development of a hypothetical super food processor to illustrate some of the difficulties in developing a product.

CHAPTER 9

Achieving Overlapping Activities

This chapter covers a core technique for saving development time. So far we have covered the savings available from getting an aggressive start at the front end of a project, creating a set of limited product objectives, and shortening the decision-making loops through techniques of staffing and project structure. Here we discuss the powerful opportunities that exist to start activities even before their traditional predecessor activities have been completed. Then we examine how to work in parallel on tasks that competitors are still completing sequentially. The overlapping of these various activities provides opportunities to make dramatic reductions in development time.

This chapter is built on the preceding ones and goes on to others that discuss some of the tools that must be used in order to put the methodology of overlapping into play without creating a mushrooming mess. The chapters following this one thus deal with subjects such as managing risk, monitoring progress, allocating resources, and areas that need special top-management attention. Managing the development process differently will let overlapping work effectively and thereby reduce development time.

This chapter is also important because overlapping is such a powerful technique that it provides an excellent opportunity to create a durable competitive advantage for a company. Many time-saving opportunities such as buying a CAD system or using a particular supplier that can provide parts quickly yield only a small edge in the marketplace because competitors can easily copy them. However, changes in product development procedures are custom tailored to an individual company. They mesh with the inner workings of a company and thus cannot be copied easily. It does take some work to build individualized overlapping methods, but this work will enjoy some protection because it will not be of much value to another company.

It can, in fact, be counterproductive to try to use someone else's product development process. At a small electronics company we visited recently the R&D manager proudly showed us their new development process, which they had taken directly from Hewlett-Packard. He reasoned that since his company makes test instruments, as does Hewlett-Packard, their proven process was appropriate for his own company. But just because of its size this hundred-employee company did not need all the complex baggage that Hewlett-Packard does to run a large, diversified company. HP's product development process would smother this company and destroy the competitive advantage that its smallness provides. The moral is that as enviable as HP's record may be, don't use their development process, because it will not fit your needs.

FROM PHASES TO CONTINUOUS FLOW

The product development process used in many North American and European companies is patterned after the phased project planning (PPP) process. In this process a project passes through checkpoints sequentially to ensure that all the items required by that checkpoint are in good order. Any problems are corrected and the project proceeds sequentially to the next checkpoint.

The PPP process was developed by NASA to manage massive aerospace projects. In such projects billions of taxpayers' dollars are at stake and tens of thousands of engineers in hundreds of organizations are working on small parts of the project, so PPP-type systems are necessary and appropriate. Some companies, however, have adopted the PPP process for all their development projects large or small because it provides a convenient means of managing certain aspects of risk. Finances are then not committed to the next step until all questions from the previous one have been resolved. Unfortunately, this approach altogether ignores the risk of missing the market opportunity. As is seen in Chapter 12 in detail, the PPP approach concentrates on technical risk to the exclusion of market risk.

The PPP process was designed for the megaproject. Throughout this book we have stressed the importance of avoiding having megaprojects by breaking large projects down into a sequence of smaller ones through incremental innovation (Chapter 4) and by making product architectural choices that divide a project into relatively independent modules (Chapter 6). Avoiding megaprojects allows for managing the process by using simpler, faster techniques such as overlapping.

The key feature of the PPP process is its phase reviews, which go by various names such as tollgates in specific companies. The essential difficulty with the PPP approach is that it treats a project as though it were one monolithic system either ready or not ready to go on to the next phase. In reality, different pieces of the system may be ready to go on to the next phase and might in fact benefit from doing so. The PPP approach unfortunately forces the entire system to proceed at the pace of its slowest element. In essence, having tollgates forces all traffic to come to a halt at the end of one phase before anything can start for the next phase. This makes it impossible to overlap phases. In our experience phase reviews invariably force projects into nonoverlapping segments and slow the product's development.

Figure 9-1 illustrates how the PPP process works. This diagram, which applies to an electronic instrument product, includes just the engineering (excluding software development) and manufacturing activities. Down the middle of the chart are the phases written in the common flowchart format. (This format itself suggests delay while the flowchart loops back to obtain phase approval.) Several observations should be made. One is that activities have to be completed during a given phase so that they can be reviewed. Only then are all the results from that phase available in final form to start the next phase. Manufacturing and engineering tend to work in isolation on their tasks during a phase. Most importantly, manufacturing's activities do not even start until Phase Three out of the five phases.

Contrast this process with the same activities laid out in so-called truss diagram style in Figure 9-2. Named after the arrangement of members in a structural truss, this diagram not only includes strong linkages between activities in the same function but also makes frequent diagonal information exchanges between functions, to tie activities together and reinforce the whole. Unlike the flowchart diagram, these activities do not necessarily have to be completed within their boxes, but there is enough information available so that other functions can begin working. For example, when engineering has a rough idea of the product's specification, manufacturing can take that to initiate discussions with suppliers who might be able to provide various needed product functions. (In Chapter 5 we suggested that marketing, engineering, manufacturing, and others should write the specification jointly, but here, to

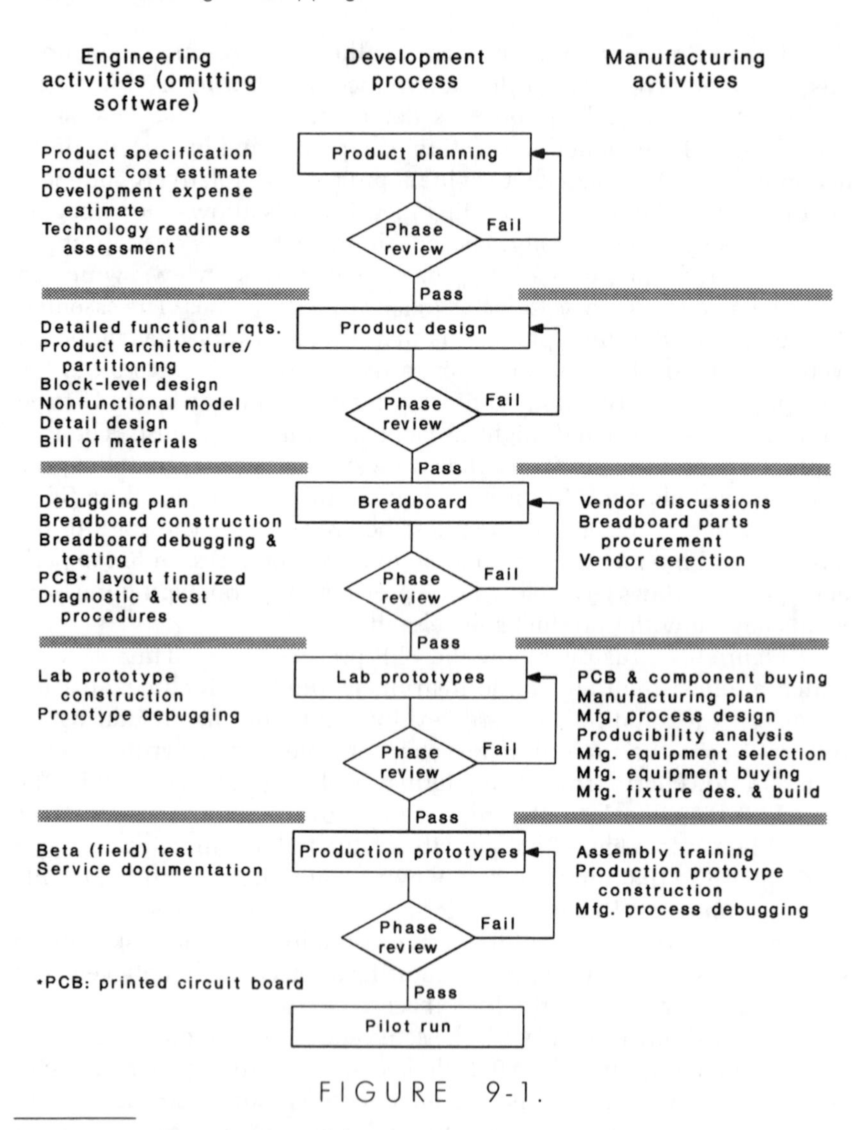

FIGURE 9-1.

Engineering and manufacturing as they operate under a sequential PPP process, which stresses project phases (an electronic instrument product is illustrated).

highlight the difference between a phase-based process and overlapping, let us simply indicate that engineering is to write the specification.)

Notice that the phase divisions in Figure 9-1 have disappeared in Figure 9-2, because the work is now occurring more continuously.

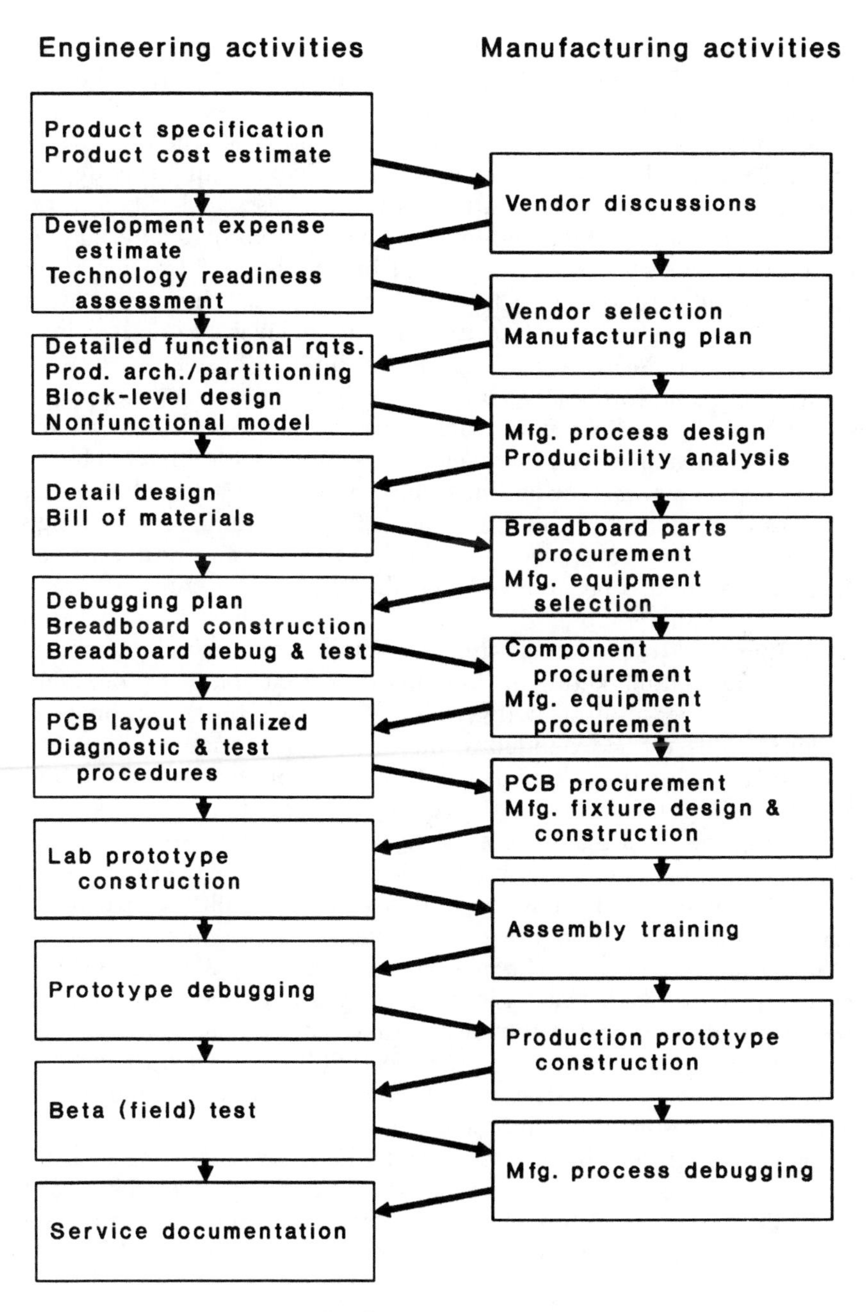

FIGURE 9-2.

The same activities as in Figure 9-1 have been organized now using a truss diagram to highlight the opportunities for overlapping.

What is not so apparent from this diagram is that everyone has begun operating much earlier, though with incomplete information. For example, manufacturing can discuss new technology needs with suppliers even before having a complete specification. They then feed these incomplete supplier findings back to engineering so that this department can firm up the specification more by assessing the state of the art of the technology that suppliers may be able to provide. Based then on a somewhat firmer spec, manufacturing can go back to some of the suppliers and go into greater detail with them so that they can contribute to relevant product architecture and block-level design decisions. In this way the design evolves as the information builds, by weaving its way back and forth between engineering and manufacturing.

Observe that although the same activities occur in Figures 9-1 and 9-2, the arrangement of the development process is different. In the truss diagram manufacturing starts much earlier and its involvement is more uniform over the duration of the project. Similarly, engineering's involvement is more constant. Such other functions as marketing and finance would also be involved, but they have been omitted here to keep the diagram simple.

Figure 3-5 was another, simpler truss diagram, which connected engineering and marketing during the fuzzy front end of a project. Although the diagram itself will get more confusing, the visual image of a truss can easily be extended to include several functions. Consider the type of square truss often used for the boom of a construction crane. Think of the four corners of the square cross-section as representing marketing, engineering, manufacturing, and finance, for example, with all the functions interlinked by the diagonal cross-members.

Throughout this book we have suggested that the difference between traditional product development and accelerated development is a matter of management approach. The most concrete manifestation of this difference is perhaps that between Figures 9-1 and 9-2. The great advantage of the PPP method is that it provides clear points for management to get involved and make decisions on a project. PPP is a system designed to give management a great deal of control, particularly when a project hits a tollgate. It does provide considerable control, and at only a small investment of management time. These points disappear, however, from the truss diagram. Instead the team is more in control of the project's destiny and management needs a new way of interacting with the team to assess its progress. Chapter 10 addresses this new approach to control.

Overlapping and Partial Information

Product development is a process of gradually building up a body of information until it eventually provides a complete formula for manu-

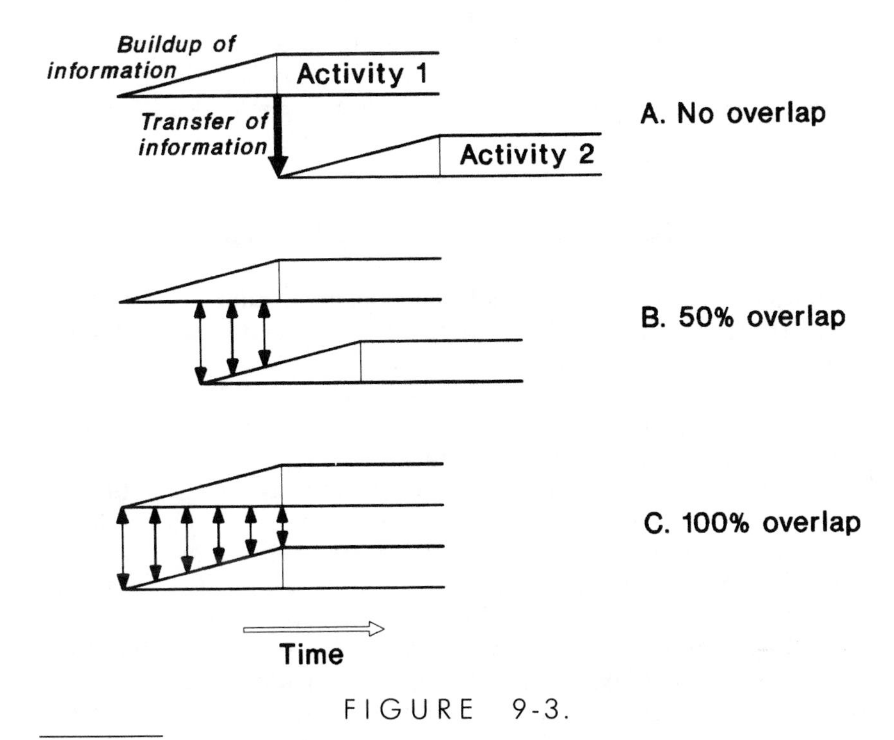

FIGURE 9-3.

Overlapped and nonoverlapped activities, showing their differences in the information-transfer process.

facturing a new product. Overlapping affects how this information is accumulated. In the traditional, more comfortable approach, information about a topic builds up until it is virtually complete, then is transferred to the next activity, where it is used to build the body of information needed for that task. Figure 9-3A illustrates this gradual accumulation and lump transfer of information. Notice that since the modus operandi is to provide complete information, the second activity cannot start until the first one has been completed.

In Figures 9-3B and 9-3C the two activities overlap. Because the first activity is still incomplete when the second one starts, the information available to it is by necessity incomplete. Working with this partial information requires a completely different style, as indicated by the difference in the arrows in Figures 9-3A and 9-3B. In the former case the information transfers in a single large piece, in just one direction. In contrast, with overlapping the information is transferred in small batches as it evolves. Because the information is incomplete, communication must go both ways as recipients ask questions about the data to find out what it means and provide feedback as to how well it meets their current needs.

Figures 9-3B and 9-3C differ only in their degree of overlapping. The amount of overlapping possible in a given case depends on how early the first activity can provide enough information for the second one to get started.

Formal and Informal Overlapping

Sometimes overlapping is already part of the standard way of doing business. A firm decides, for example, that although it sometimes creates problems to send informal drawings to suppliers for quotes, it saves enough time in most cases to make the practice worthwhile. Consequently, by assuming an acceptable degree of risk the company overlaps the bidding activity with the final drawing activity. Many other instances of overlapping have become established practice in firms when they are used routinely. Some examples are making a layout drawing before the product specification is approved, or making the first production units with makeshift tooling or by hand assembly while the final tooling or assembly machinery is still being prepared.

There are many creative ways to provide overlap in the design process. These two examples both involve machined castings. In the first case, which comes from one of our clients, the castings used as a machine base were relatively large, complex parts with long lead times that therefore established the project's critical path. The machine base is the last part the designer wants to release, however, because it is the part to which everything else attaches and is thus the most susceptible to change. This firm's solution was to design the raw casting by providing surplus material in the attachment areas. The casting drawings were sent off to the foundry to have initial castings poured while the designer refined the fit of the mating parts and completed the details of machining the casting. In essence, these cast parts were being designed even as they were being poured. (After the product was in production, its drawings and pattern could be modified if necessary to minimize the machining of surplus material.)

Another example is provided by Neles-Jamesbury, which manufactures quarter-turn valves used in chemical plants, pulp mills, oil refineries, and similar demanding applications. Neles-Jamesbury requires high-quality castings, and often the first batch or two to come from the foundry for a new design have dimensional or porosity defects. The normal practice would be to return these defective parts and wait for the foundry to make corrections and send another batch. This company found, however, that once they started machining the castings they then faced another round of learning in debugging the machining process. So they now accept and pay for the initial defective castings and use them to debug machining operations. Thus, by machining bad castings and scrapping them, Neles-Jamesbury is able to overlap two learning processes.

These examples show formal overlapping procedures that are the established means of doing business in their companies. In addition there are informal opportunities for overlapping that arise every day as special circumstances appear in a project. For instance, a team leader may need to get several signatures quickly on an engineering change notice (ECN). Rather than working sequentially to gain approval, he or she decides to send an unsigned draft ECN (the partial information) to all approvers simultaneously to initiate discussion. Then the actual ECN can be approved quickly. This process was devised for one-time use to fit special circumstances.

A development team can spot such individual opportunities and exploit them if they are in close communication about the project's objectives and are creative and willing to experiment. Management support is critical to this type of experimentation. Such ad hoc overlapping and the type of continual emphasis on saving time that supports it are the breeding ground for the formal overlapping processes that ultimately build competitive advantage.

CREATING OPPORTUNITIES FOR OVERLAPPING

Fortunately, there are several things that can be done to find techniques of overlapping appropriate for a given company. First recognize that the use of dedicated project team members discussed in the previous chapter works to encourage overlapping. Think about how a project is usually staffed. In its early phases marketing may provide most of the effort, then engineering gets heavily involved, and finally manufacturing becomes the dominant player, as illustrated in Figure 9-4.

Now consider a team staffed from Day One with full-time members from various functions, as suggested in the last chapter. See Figure 9-5, which should be contrasted with Figure 9-4. The part-time, phase-in, phase-out type of staffing in Figure 9-4 encourages—or rather condones—a sequential approach in which engineering tends to follow marketing and manufacturing takes over after engineering. In Figure 9-5 different functions are *forced* to act in parallel, with manufacturing forced to be involved from the beginning. If the manufacturing people join the team full time from the beginning they will either have to work on engineering or marketing tasks or identify opportunities for overlap

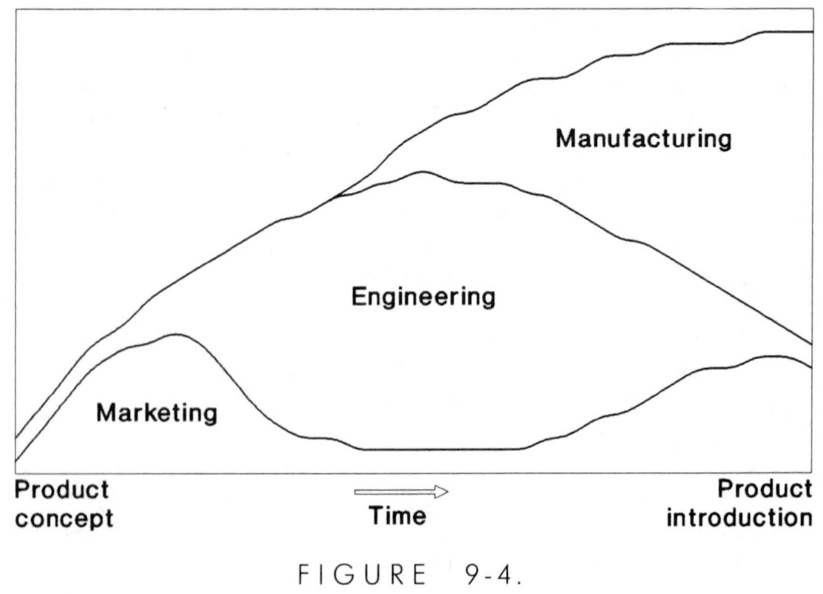

FIGURE 9-4.

Conventional staffing practices force development efforts into phases that become dominated by specific functions.

so they can begin work on manufacturing tasks, even if this must be done with only partial information.

Overlapping relies heavily on the use of partial information. The effective use of partial information both requires and supports the concept of a close-knit team. Lots of face-to-face communication is essential, so it is beneficial to have the participants close together, or co-located in our term. Partial information is too fragmentary and dependent on feedback to rely on getting it from intermediaries, so principals have to be talking to principals.

The search for ways to get started earlier with partial information encourages a problem-solving orientation, stimulates initiative taking, and tends to keep the focus on the overall project goals. Because everyone is working with changing information, the group will naturally develop a sense of cooperation and of sharing the responsibility for the outcome. Simply put, the team has to work together in order to succeed.

Getting into the Overlapping Mode

Just getting the full-time team members together in one location does a lot to foster an overlapping approach, but a catalyst may be needed

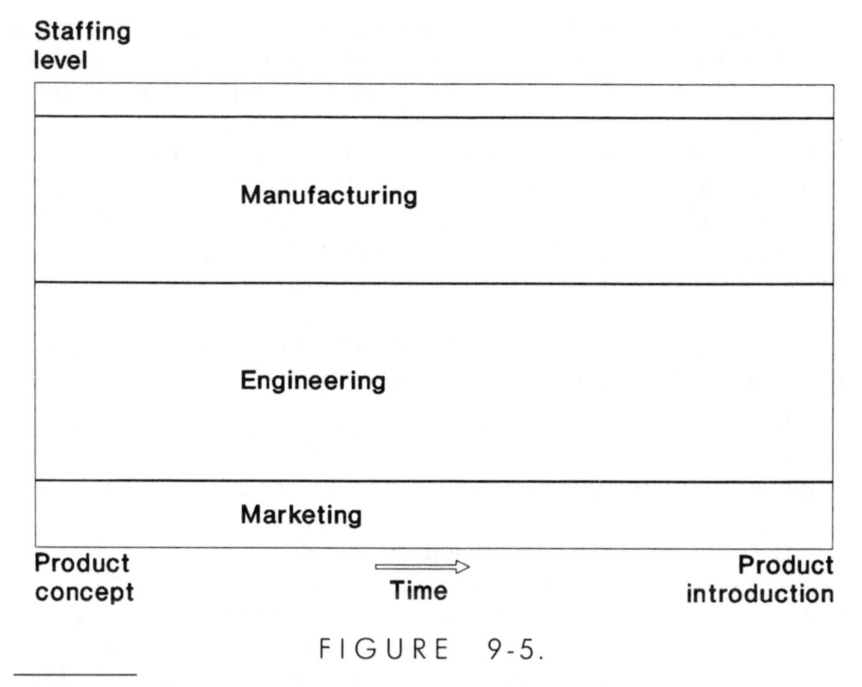

FIGURE 9-5.

Dedicated staffing forces overlapping and cross-functional interaction that uses partial information.

to get the reaction started. In general, the team members should be the ones to create the overlapping. This process requires a keen sense of the data that must be transmitted from one activity to another to support the actual sequence of design decisions that must be made in the later of the two activities. Since this requires exact knowledge of the sequence of design decisions, which is a notoriously foggy path, it is usually best identified by the people performing the two activities. They are closest to the problem and will be likely to have the greatest understanding of it. They will have to execute the overlap plan, so it is much more likely to succeed if they create the plan and feel they "own" it.

On the other hand, being too close to the problem can also get in the way. If team members are too close to the problem to see radical alternatives, introduce an outsider who can ask a few "dumb" questions to get the creative juices flowing again.

One of the best ways to find possibilities for overlapping is to work backward. Start with the desired end point and see what has to be done to arrive at it. This may even suggest an "unnatural" ordering of activities.

Look for examples where something got done exceptionally fast, even if it was only once. On one consulting assignment we were assisting a task force trying to cut part of their development process down from thirteen months to three. The task seemed hopeless to those old-timers, and we weren't making much progress. Then a respected member of the group noticed that, once before, they had done the job in only four months. By studying this example they soon had a plan for routinely doing the job in four months. Mathematicians often do their work by starting with an "existence proof," first demonstrating that there is in fact a solution without worrying yet about its nature. Sometimes some of the rest of us need an existence proof to spur us into finding a solution.

More specifically, look for opportunities for overlapping by asking some questions about partial information:

- What is the bare minimum of information needed to get the next step started?
- When is the earliest I can produce this information in the preceding step?
- Is there anything that can be done to make the requirement for having this information unnecessary?
- Are there assumptions I can make about this information that will provide a high likelihood of its being accurate enough to begin work?
- Are the likely consequences of a particular mistake going to be large or small?
- Can one save enough time by starting early to allow for making a mistake and still finish early?
- What information would allow taking another step?
- Who could use the information I have to enable them to take another step?

Overlapping is done both by taking an overview to look for major opportunities and a microscopic look to find ways of inching forward with the information at hand.

"Pulling" Partial Information

Overlapping amounts to more than just doing things in parallel. Any competent project planner can look at a project and see activities that can be executed simultaneously, then schedule them accordingly. That sort of parallel processing is obvious and you are probably already taking advantage of it. What else can be done?

As indicated, taking the next step requires employing partial information. The just-in-time (JIT) manufacturing philosophy can help us here. JIT has come up several times throughout this book because it is a powerful approach that has helped many manufacturing companies become more competitive. JIT is to some extent the manufacturing analog of the kind of improvements we are trying to make in the product development process. A great deal of effort has been devoted to refining JIT, and some of its techniques now relate strongly to what we are trying to accomplish in accelerated product development.

In the present context let us relate the "pull" concept of JIT to the use of partial information. In traditional manufacturing a batch of parts is made at one work center and "pushed" to a downstream work center whether or not the recipient is ready to use the parts. In a pull (sometimes called *kanban*) system the downstream people call for the parts when they are ready for them. It is said that the pull concept was conceived by Japanese manufacturers who had observed how an American supermarket works. There customers control the flow of goods by pulling items off the shelf as they need them. The store simply restocks the shelves as they are emptied.

Traditional product development, in the PPP model, is a push system. It operates as shown in Figure 9-3A where a large batch of complete information is pushed on to a downstream activity at infrequent intervals. This information may not all be useful, but even if it is it will take a while, as indicated in the figure, for the downstream activity to absorb it.

As illustrated in Figures 9-3B and 9-3C, partial information could be either a push or a pull system. There are advantages in encouraging a pull approach, however. For one thing, if the information is pulled it will come sooner and be in smaller batches. It will also likely be more applicable and more recipient oriented. There will be less chance also that the upstream activity will waste effort on something that is not needed.

The pull approach is established on a development team by making it clear that it is the responsibility of the downstream person to ask for whatever information they need. By making this the standard operating procedure the downstream tasks will naturally get started sooner, compressing the whole development cycle.

There is another benefit to the pull approach that becomes evident as a firm starts using dedicated teams to develop its products. Observe that in Figure 9-1 manufacturing apparently had nothing to do at the beginning of the project. We know that there are things that the manufacturing people should be doing, but someone must discover what those are. The pull approach places the responsibility of finding them squarely on manufacturing. By making it the responsibility of the receiving party to ask for the information they need to do their job, they will automatically have something to do on Day One. It should be left up to them to take the initiative for their part of the job and ask for the information they need to complete it. Unless they do, nothing will happen in their area for a while because all the upstream people will have plenty to do without concerning themselves about downstream activities.

Moving to the pull system just described will be a substantial change for many companies in which the push system is deeply ingrained. In our consulting we are often told by manufacturing people how overloaded they are by all the projects pushed on them by engineering, and engineering tells us about the heavy load dumped on them by marketing. Some forward-thinking companies have cut the inventory and lead times on their factory floor through using JIT pull systems, but they have yet to apply JIT pull techniques to their product development processes.

New Skills and Attitudes Needed

Effective overlapping depends on the adept use of partial information. However, because partial information is difficult to work with, some special skills must be developed. One especially critical skill is the ability to project the possibilities of what might happen if the partial information turns out to be wrong. The recipients of the information in particular must develop a sense of just how far they can go with the information they are given without going too far astray. It must be recognized that they will go astray occasionally because if they don't they aren't pushing—or rather pulling—hard enough. The recipients of information must constantly be projecting what its likely outcome will be and what some possible alternatives are. Obviously, the more frequently they talk with the providers of the information, the better their forecasts will be.

The providers of the information also have an obligation to try to appreciate where it may lead the recipients. They need to be aware of how the information will be used and be sensitive to the impact on downstream activities of any changes they may make. They also need to keep in touch constantly with those working with their partial information.

It should be clear by this point why the organization needs to provide the best possible project structure to enhance communication for those who will be using partial information. Without close communication, the loose ends created by working from partial information can create havoc.

Underlying the new communication and forecasting skills being developed is a set of attitudes that will ultimately have to change. Chapter 15 discusses the changes in attitudes and behavior that come about in making the transition to accelerated product development. Here let us simply stress that the prevalent attitudes in many companies are inconsistent with working successfully from partial information. The common operating style—which is almost a game in some companies—is for each person to wait until everyone else has completed their part of the project and only then start, after nothing can change to mess up their work. Nobody wants to discard work that somebody else's actions have made obsolete. Thus, to protect themselves people try to start as late rather than as early as possible.

Beyond this there is a general distrust of upstream information. Manufacturing believes that engineering will forever be changing its drawings, and engineering is uncomfortable with the information coming in from marketing. In a recent survey in *Research-Technology Management* (November–December 1988, p. 39), 72 percent of the R&D managers questioned stated that they do not use marketing information because it is incomplete, and 49 percent don't use it because they find it inaccurate. Logically, this may seem silly, because incomplete information is still better than no information. Nevertheless, it illustrates the pattern in which people are reluctant to take any sort of action until all the facts are in.

Before engineering is to start requesting and using the even more fragmentary marketing information, their perception of its value will have to change. Engineering will have to recognize that having partial information is better than none and that the consequences of waiting for "perfect" information may be far more severe than moving forward with imperfect data. Risks must be taken in using this partial information, but they can be managed.

SUGGESTED READING

Clark, Kim and Takahiro Fujimoto. 1989. Reducing the time to market: the case of the world auto industry. *Design Management Journal* 1(1): 49–57. Clark's work focuses on the overlapping of activities and the use of partial information in automobile development. Interesting details on overlapping of the design and fabrication of stamping dies for body panels, which is a complex, interactive process.

Hay, Edward J. 1988. *The Just-In-Time Breakthrough: Implementing the New Manufacturing Basics*. New York: John Wiley & Sons. Of the many books available on JIT, this is a recent, good one. Just as JIT has revitalized manufacturing, the product development process is in need of revitalization. Some proven JIT concepts carry over to accelerated product development, as has been illustrated in this chapter. Product developers can learn much from JIT.

Takeuchi, Hirotaka and Ikujiro Nonaka. 1986. The new new product development game. *Harvard Business Review* 64(1): 137–46. Perhaps the best piece available on techniques for accelerating product development. Includes a good section on overlapping that draws on examples from Honda, Canon, and Fuji-Xerox.

CHAPTER 10

Monitoring and Controlling Progress

Controlling the progress of activities is one of management's basic duties, which certainly applies in a product development project. In particular the R&D portions of these projects are regarded as especially difficult to control, however. Difficulties arise out of the inherently unpredictable nature of the problem-solving process. It simply is difficult to predict the amount of time needed to solve an unknown problem and how adequate the solution will be when it does arrive. The prototype of a new product, for example, may not be working today, and it provides no information as to how much debugging will be needed before it will work.

Before delving deeper into the uncertainties of the development process, consider the unique nature of fast development projects. Accelerated product development is even more difficult to control than that of conventional products, for two reasons. First, in a fast project the development process is more complex, because there will be overlapping activities. It is a twelve-ring circus, not a linear progression of events. Second, because most of this work is done with partial information, redirection of a project can occur as new information emerges.

These difficulties make conventional methods of identifying problems, tracking progress, and making corrections inadequate. They are neither fast enough nor precise enough, and often they introduce delays of their own, as for instance delays in convening a group to review the project. A different approach is required for managing the progress of an accelerated project.

This new approach requires a fundamental shift in management attitude away from controlling and toward empowering and enabling the success of the team. Only by using team members to control their own progress can a truly rapid process be created. The staffing and structural provisions in chapters 7 and 8 are aimed at creating a team with the ability and desire to control itself. The team itself has more precise information upon which to base decisions than does management, which argues for its being endowed with the desire and authority to strive for its own success. When control rests with management rather than the team, the desire for success naturally tends to shift toward management. Then the decisions are made by those who do not have all the facts, and decision making becomes too slow and infrequent to maintain a rapid development pace.

THE PRODUCT DEVELOPMENT PROCESS

The nature of the development process must be understood before it can be controlled. Sometimes the process, and in particular its engineering portions, seems rather mysterious to those who have never been directly involved on a development team.

Marketing information may be less intimidating than engineering information, but it is also generally regarded as being soft. This degree of uncertainty tends to frustrate engineers, who would like marketing's research to reveal exactly how big an instrument's case can be or just how much fan noise is objectionable.

We tend to infer from evidence like this and from the fact that engineers are always making calculations that engineering is a precise science. In some ways it is. Engineers learn how to calculate precisely how much a cantilever beam will deflect or what the loss will be over a transmission line. The pitfall in this logic, which is rarely explained, is that there are no ideal cantilevers or transmission lines in the real world. Engineers have to start making guesses as to how unideal their solutions actually are. The precise techniques that comprise the bulk of their academic training are ultimately of only limited use in actual design problems. When established techniques fail to apply, the process quickly becomes more empirical than analytical.

Engineering professor Billy Vaughn Koen, of the University of Texas, is noted for his research on the theory of engineering design. In a monograph, *Definition of the Engineering Method* (American Society for Engineering Education, 1985), he provided his definition of the engineering method:

> The engineering method [is defined] as the use of engineering heuristics to cause the best change in a poorly understood situation within the available resources. This definition is not meant to imply that the engineer just uses heuristics from time to time to aid his work, as might be said of the mathematician. Instead my thesis is that the *engineering strategy for causing desirable change in an unknown situation within the available resources* and the *use of heuristics* [Koen's emphases] is an absolute identity. In other words, everything the engineer does in his role as engineer is under the control of a heuristic. Engineering has no hint of the absolute, the deterministic, the guaranteed, the true. Instead it fairly reeks of the uncertain, the provisional and the doubtful. The engineer instinctively recognizes this and calls his ad hoc method "doing the best you can with what you've got," "finding a seat-of-the-pants solution," or just "muddling through."

If "muddling through" is an apt characterization, this has consequences for controlling progress. One is that it is both difficult and impractical to control progress at the microscopic level, where a design's progress looks pretty chaotic. Here it is two steps forward, one back, another sideways, and finally a hop to avoid an obstacle. Trying to use elaborate, highly quantitative control methods on a seemingly random process is wasteful and likely to be frustrating. Choosing competent, motivated, seasoned team members is therefore the best control strategy in a situation like this.

The other consequence is that the end point itself can get muddled while one is muddling through. An acceptable end point needs to be kept clearly in team members' minds, and especially in the leader's. He or she has to be able to see when a certain part of a product is good enough. Then it can be deemed finished and the people working on it be reassigned to something that then really needs their help. This type of control and guidance is done much more effectively by a team leader than by outside management.

THE PRIMACY OF THE SCHEDULE

The critical management controls are those that monitor the schedule as through a series of tripwires, alerting management when to come to the aid of the team. Some corporate control systems such as progress reviews are helpful to use in maintaining schedule progress, but many others, like travel approvals, serve more to impede progress. Used to excess, such cost-oriented controls become bureaucracy. The solution to bureaucracy is not to remove all controls but to temper them so that they do not slow up a project. The question then becomes which controls must be tempered and which be kept rigid.

The answer is driven by the economics of the development process. Usually, having performed the type of economic analysis suggested in Chapter 2 for a development project, we find that a development delay will have a much more serious effect on profitability than would the development cost. In this situation we must appropriately strengthen controls that relate to the schedule while streamlining cost-related controls, particularly ones that might slow the project. For example, we must have the tools to determine whether a supplier will deliver parts on time, but we cannot tolerate a system that requires four weeks to approve a visit to a supplier.

This is not a simple case of discarding all cost and performance controls and making the schedule the king. Rather, scale back the controls so as to have tight control over the cost of the $1 million tooling outlay but only simple controls on the use of petty cash by team members. The biggest mistake we see as consultants is excessive use of accounting-driven, cost-oriented controls. Such controls should be vigorously modified whenever they threaten to delay development.

One solution is to design appropriate flexibility into the controls. For example, an engineering manager at one company explained that he would approve travel, even on chartered aircraft, for any number of his people without justification if they wanted to travel to another company facility. His company has several facilities, and he knows that separation hampers communication, which in turn slows progress. Besides, the plants are in locations where it is not likely that the travel privilege will be abused. For other destinations the normal travel-justification procedures apply. The point illustrated is to soften only that aspect of a cost-oriented control that influences the schedule.

Corporate capital-authorization processes are frequently cited by team members as reason for delay in projects. Yet we have found that sometimes this issue is a red herring of making distant corporate accountants become convenient scapegoats. In other cases upon investigating it is clear that the approval process is indeed slow, as at one firm that took an average of nine months to obtain approval to order tooling. In the smoother-running operations capital authorization is not a sequential step that catches financial planners by surprise. Instead, it is a parallel, carefully integrated activity that develops along with a project's design and is prepared to supply funds as needed. Big expenditures are still carefully controlled, but the development process never stops for lack of funds or approval.

Securing the capital needed for a product development project often requires a highly visible, significant decision, but the greater cause of delay seems to be the controls on the hundreds of seemingly insignificant financial decisions. For example, how liberal is company policy about dispensing petty cash to an engineer who is working on a Sunday and has to buy some parts at a hardware store? The company's exposure here is minimal, and the effect on the schedule may look small. Sometimes, however, just one day can make the difference between getting a drawing to a supplier on Friday to get it into the weekend work schedule or having the supplier start work sometime the next week. Furthermore, the true damage of some of these apparently inconsequential control systems lies in the underlying message they convey about corporate values. If an engineer's reimbursement is challenged, that is a powerful message from the system about the relative importance of following the accounting rules as against exercising initiative and trying to design the product quickly. Chapter 14 covers this subject in greater depth.

SOFT CONTROLS

Beyond the distinction between schedule-oriented and cost-oriented controls there is a difference between hard and soft controls. Soft controls are the relatively subtle ways there are of empowering and motivating people to work effectively in meeting project goals. Hard controls are simply variants on the familiar types of product development procedures, project planning, and project reviews. Soft controls are discussed first because, in addition to their being directly useful, they provide an appropriate context for viewing hard controls.

Effective project control starts with product definition (Chapter 5) and staffing (Chapter 7) to establish a shared vision of the product and build ownership in the concept. To the extent that each team member understands what the project is and accepts responsibility for completing the job as a whole, the individuals' efforts will be applied more effectively. By contrast, hard controls such as formal project reviews are by nature reactive and thus less effective, especially on a rapidly moving project. To borrow a phrase from the manufacturing quality movement, the objective in product development is to build quality in, not inspect it in. In this context formal design reviews imply rework (looping), which is slow and unpredictable.

This does not mean that fast development projects are not reviewed. In fact, they usually receive far more reviewing than do traditional projects, but the process does not seem like review. Most of the reviewing is done either by peers or the project leader, informally and frequently. Right after one such review if you were to ask the reviewee if he or she had just had a review he or she might say, "No, but I'm sure glad Jack stopped by and mentioned that our molding supplier usually shuts down for the first two weeks of July, because that's just when I was planning to get my mold changes to them. I guess I'd better give the supplier a call to see what we can work out while they are shut down. Oh yes, Jack also thought it might be a good idea to check with the resin supplier to see if the new compound they are providing needs an adjustment in the fillet radius before I finalize things with the mold designer."

In accelerated projects team leaders spend a lot of time "chatting" with members of the team to keep up to date on progress and problem areas and make sure activities dovetail smoothly. If a subject gets outside the leader's limits of competence, the leader should arrange for a person experienced in that area to "stop by and chat" with the team member, as in the case above regarding the mold. Such informal conversation is particularly essential when the team member is relatively inexperienced in an area, especially when that person is too new to the company to have developed a peer network yet. The coaching role, as distinct from the directing one, comes naturally to some team leaders, but others have to develop this skill consciously.

The staffing and organizing principles covered in chapters 7 and 8 enable the softer or more subtle forms of control to be effective. Basically, the technique is to put the team into a common environment and provide them with a common objective and common reward. Doing so encourages the open communication and trust needed to share partial information effectively and reliably. Partial information is by its very nature earlier information, so with it the team starts performing tasks sooner and can move more confidently with its partial information. Decisions get made faster because all parties to the decision have the same information and are working toward the same goal. The control in this situation really stems from both self-motivation and peer pressure generated by intense daily interaction as the group works toward a common goal. These relationships are illustrated in Figure 10-1.

Frequent, Open Communication

The kind of communication just discussed is frequent and open, but much corporate communication is not so fast. Memoranda in particular

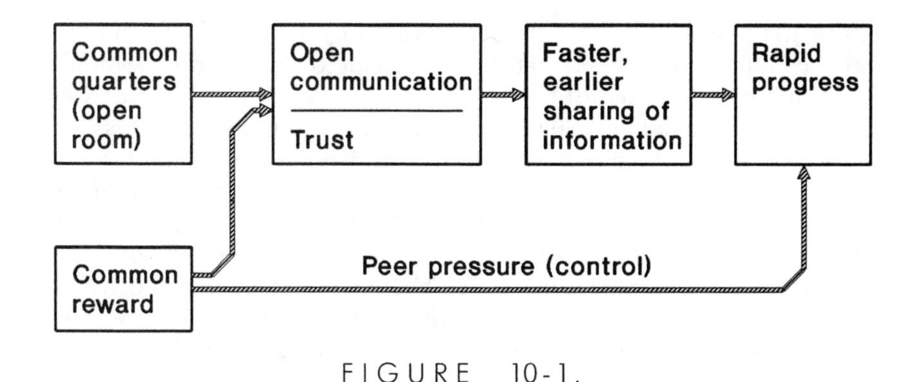

FIGURE 10-1.

The effect of sharing common quarters and a common reward in accelerating development progress.

are slow. It takes a long time on a scale of minutes or even seconds to compose, produce, and distribute memoranda. This is where the biggest delay usually rests, because a memorandum may not be read for days, and it may never get a response.

Electronic mail speeds up some parts of the process, but it still does not get around the three fundamental flaws of written communication. One flaw is that the very act of composing a written message can unwittingly draw a person into solidifying a position, advocating a particular sequence of events that will involve others, and making assumptions about how a reader of it will respond. None of this helps to keep communication open. In contrast, in verbal communication a position is taken progressively as a conversation is shaped by the reactions of the person with whom one is conversing. Second, written communication can easily go into inappropriate detail or neglect important information, because it lacks the feedback of an audience saying, "I know that" or "I don't understand." Written communication invariably provides both too much and too little information for its audience. Finally, the timing of a response to written communication is unpredictable. The recipient may call immediately and share just what he or she thinks of the memo, but if there is no response, what can you assume? Had you delivered the message in person and the recipient had stormed out of the room without saying a word, there would at least be some information about the reception the message got.

Regular, brief written reports to management can be used to maintain their enthusiasm for a project and forewarn them of any difficulties the team is having (executives don't like to be caught off guard about the status of one of their projects). Even here, though, the technique of writing is not preferred. Rather, management should receive most of its information—faster and more interactively—by visiting the team regularly. (Chapter 14 covers this "management by wandering around" technique in greater depth.)

The team needs to have its own regular meetings to discuss the project's status and share new information that may affect the project. These meetings may also be appropriately used for problem solving if the problem is one that affects most of the participants. If it does not, however, the problem is best handled at a smaller meeting so that it does not waste other people's time. The beauty of having the whole team co-located in one open room is that people will invite themselves to a problem-solving session just by joining the discussion if they think it may affect their part of the design.

Regular team meetings can be handled in various ways, depending on your style. The important point is to hold them frequently but to keep them short and informal.

Here is how this worked out in practice in a 100-person company in the Boston area that manufactures computer equipment. Because they had had difficulties with design engineers going too far afield and having to redesign a product when others thought the design had gone off on a tangent, they tried daily team meetings. Every day at the ten o'clock coffee break (which encouraged informality and tied the meeting to something pleasant and regular), the team met to discuss the progress made in the past twenty-four hours and set goals for the next day. The group included quality control, manufacturing, purchasing, field service, and often marketing. With all these key players involved frequently, nobody could get very far off the track, and the design engineers' time was therefore spent much more productively. They were also more open to making changes while their ideas were still fluid. Designers were not then second guessed after days or weeks of work, which would have destroyed their morale. Moreover, the decisions about the project were being made at the lowest possible level, on an open and cooperative basis by those who would actually have to execute them. Purchases were approved on the spot and similar administrative issues resolved immediately.

There is another interesting feature about the way this firm handled team meetings. The meetings were set up by the company's vice president, who attended them whenever he was in the plant. He did not really run these meetings, but he did make sure that recent developments were discussed and plans made for the next twenty-four hours. Although he was dedicating precious time to attending these meetings, there were two big benefits from it. Management continuously knew a great deal about how the project was going, and his regular presence in meetings sent a strong message to the team about the importance of the project.

There is always room for improvement, however. In retrospect, this company found that it had not gone far enough. In this project they had isolated a key subsystem and given it to an outside consultant, according to the principles in Chapter 6. Then, to provide continuity, one of the company's engineers interfaced with this consultant and reported on his progress to the daily meetings. The consultant's work was good and on schedule, but the team eventually learned that the function provided by the subsystem could have been done in hardware rather than software, at a lower cost and in significantly less time. It turned out that the liaison engineer was unconsciously filtering the information he was reporting back to the team, with the result that decisions were made that would have been different if the consultant had participated directly in team meetings. It was resolved that next time the consultant would be a direct participant.

The daily meeting concept just described is not limited to use in small companies. For instance, consider a large computer company with different requirements that handled its meetings differently, but still held them daily. Engineering and manufacturing were interdependent in this company, but were located fifty miles apart. Their solution was to "meet" on a conference phone call every day at four o'clock. The parties on the two ends of the conversation would vary some from day to day, and sometimes they might not have much to report, but the call would always be made. Because phone calls have their useful limits, there was also a regular weekly team meeting, always held at a fixed time but alternating in location between the lab and the plant.

The discipline of holding these daily meetings forces people into thinking about making daily progress. In a fast project every day counts and no one can afford to lose a day, either because nothing got done or because the wrong thing was done because of poor communication. Monthly or quarterly reviews simply do not do the job in rapid product development.

Finally, the team leader must set the proper tone for these frequent meetings. They are for sharing information, solving problems, setting goals, and maintaining momentum. If the daily meeting degenerates into a mode in which each member simply reports as having completed what was assigned at the last meeting, then the emphasis has shifted from team accountability to individual "cover your rear" behavior in which the team members have ceased to focus on the success of the whole project.

Communicating is admittedly difficult and demanding. Oral communication has the advantages of immediate response and higher fidelity through body language, but it still has countless pitfalls. One aid to better communication is to keep the conversation as informal as possible. In our consulting we sometimes ask engineers what they think of having the CEO wander into the lab and talk with them. They typically say that they would like to have the CEO show more interest in the new products under development, but they are uncomfortable when he or she shows up only once a year. More-frequent encounters would help keep the project environment less formal.

It is usual to think of speaking as being the more difficult part of a conversation, but high-quality listening is probably harder. If a team leader really wants to know how a project is proceeding, he or she needs to be a good and patient listener.

By their nature, product development teams tend to be heavily staffed with engineers, who will have studied lots of math but probably not much psychology. Engineers are trained to construct logically tight arguments and filter out apparently extraneous emotional factors. The problem with this approach is that fast product development depends less on scientific purity than on such human emotions as enthusiasm, satisfaction, frustration, boredom, and depression. The astute team leader thus pays attention to these factors rather than filtering them out.

HARD CONTROLS

As may be clear by now, we are not so much interested in doing away with hard controls as in making them more agile by softening them some from the way they are normally used to control projects. In addition to the hard controls covered here, Chapter 13 discusses appropriate controls for design documentation, including drawings, bills of material, and engineering change notices.

Product Development Procedures

Every company that develops products uses some kind of procedure that is intended to control progress. The procedure may be heavily documented or just stored in the team leader's head, be rigidly adhered to or quite flexible, sequential or overlapping. To an extent the type of procedure a company uses will depend on its culture, and a culture tends to change slowly unless catastrophic business conditions force rapid change.

There are, however, factors that suggest the degree of control needed, such as the size of the project. The status of a small project can at least initially be stored in the team leader's head. A medium-sized project with more than a dozen people involved needs some sort of structure to manage all its details. And a large project, like an automo-

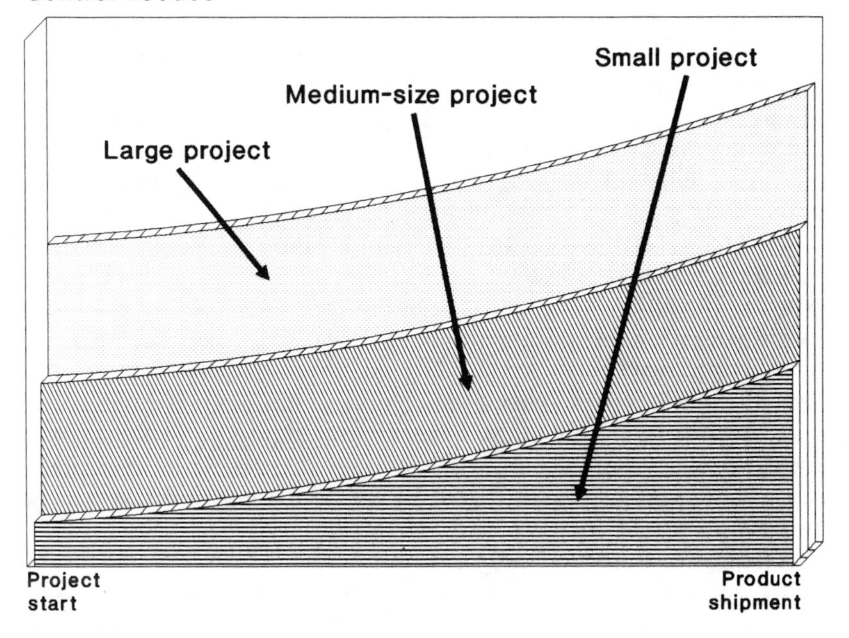

FIGURE 10-2.

The degree of project control needed depends on the size and stage of the project.

bile or mainframe computer, requires a substantial procedure just to manage design interface issues. Figure 10-2 illustrates the amounts of control needed to manage these three types of projects.

As the figure suggests, the degree of control needed also grows as a project progresses in its development cycle. As the project matures, the investment made in it and the number of people involved with it both increase, the number of details (for example, the number of design documents) mushrooms as it moves toward production, and there is not only less time available to react to problems but more extreme measures required to solve them.

Note that we have not suggested that the amount of control needed depends on the company's size. In fact, we have commonly seen big companies developing a "one size fits all" development process that overburdens small projects with bureaucracy. It is the size of the project, not of the company, that determines the degree of project control required.

Designing a rapid development process requires more than simply making it loose or rigid. In our consulting assignments we encounter companies hamstrung by a rigid formal development process that

creates considerable delays in decision making and encourages continual redirection of the project. At the same time we see companies without an established development process that are immersed in confusion, which leads to slow decision making and changed product goals. One particular large company making a standardized product, which would apparently be an ideal candidate for a fairly structured process, had no established development process at all. They depended on "tribal knowledge," as one executive explained it to us. The consequence of this informality was that team leaders ran projects according to their own styles, and others didn't appreciate the milestones or their own responsibilities, because they varied from project to project. Even more devastating and more apropos here is that management did not know how to react to a project. They neither knew what kinds of decisions were needed to move a project along nor which decisions would rechannel efforts if made late, resulting in delays and squandered design time. This company has now implemented a structured plan in which executives are presented with narrower and narrower decision choices as a project matures.

There are limits, of course, to how detailed a codified development process should be. It can easily get too complicated and lose its value. Honeywell went from a three-inch-thick development manual to a twenty-page guideline. General Motors develops automobiles by using a pocket-sized booklet of only sixty pages, including some blank ones for managers to add notes on as they carry the booklet around on the job.

Most companies tend to gravitate toward development processes that are too rigid for an accelerated development cycle. They use a standard process usually intended for their largest projects, but if they were to accelerate development by following a policy of incremental innovation, as recommended in Chapter 4, their accelerated projects would be relatively small ones. They carry excess baggage for such a short trip.

Adopting a rigid process is an easy escape from making hard choices in organizing and staffing development teams. It is an attempt to build judgment into the development process rather than tapping the judgment of team members.

We advise keeping the formal development process simple enough that everyone understands it and flexible enough to allow it to be adapted to specific situations. The process should distinguish clearly between the responsibilities delegated to the team and those retained by management. In most organizations the team should be responsible for all tactical decisions, including the product's features, performance, style, cost, and its schedule within the specified envelope. Management's responsibilities relate to more strategic issues such as the allocation of resources to the project, the decision to make a significant change in product development objectives, or a decision to cancel the project. When management micromanages a project by getting involved in design trade-offs and project problem solving its schedule suffers, and frequently the design will suffer as well.

One way of separating management's responsibilities from the team's is to list management's explicitly: adding resources, altering project objectives, and cancelling the project. All other responsibilities are then presumed to lie with the team.

A more highly structured process generally reduces technical risk but in turn generates extra queue time, increases the nondesign workload, and tends to inhibit task overlapping. This lengthens the schedule and increases market risk. (See Chapter 12 for a discussion of technical and market risk.) However, an appropriately structured procedure can shorten the schedule by highlighting critical milestones and forcing faster management decisions at those points.

Project Plans

The project plan is a central item in an accelerated development project. No project is likely to be completed any earlier than is called for in its plan. On the other hand, if the plan is unrealistically optimistic it is unlikely to be followed, with substantial slippage probable. Consequently, a project's plans have great influence on controlling its progress.

There are two steps to project planning: creating a plan, and executing it. With many types of planning there is considerable value just in creating the plan. Doing so identifies critical-path items, clarifies sequences and dependencies, and focuses attention on critical points. It is best then, of course, to try to execute the specific plan, but even if the plan is not used the step of creating one alone will inherently influence the subsequent execution.

We have found this rarely to be true in project planning, however, for there is little value to be derived from a product development plan until it is used. The plan should be created by the team, who should be prepared to execute it.

The first step in planning is to set a clear end point for the project and back the schedule up from there. As clear-cut as this may seem, it can sometimes be tricky, because people can view an end point differently. For some projects it is sufficient to say that the end point is when the first production unit rolls off the assembly line. With others, such as business equipment, the distribution of service manuals and spare parts may also be required. For consumer goods high-volume production is often critical, so the end point might not be reached until the production rate reaches a certain value. In no case should the end point of a project be delivery of a design to the manufacturing organization; the plan should always span the transition into manufacturing.

In the development projects we encounter as consultants we are often surprised at how frequently development plans are torpedoed by such simple details as having a box in which to ship the product. This is easily rectified in most cases, but in one project a special carton with its graphics, drop testing, and related tasks added an unexpected four months to the project's schedule when the organization thought it was complete.

In another case the team wanted to define the end point as the time when there was a complete pile of parts on the assembly floor, because they felt powerless to work with the union to develop assembly plans. Fortunately, management spotted this shortcoming in the plans and corrected it, but if they had not there would have been months of tail-end delay while assembly workers tried to create an assembly procedure long after the development team had departed for a new project.

Another type of end point, called the break-even time, is being used at Hewlett-Packard. By focusing on the point at which a product's early earnings just offset the cost of developing it, the HP end point captures not only the development time but also provides a measure of how well the product is satisfying marketplace needs. This comprehensive measure is ideally suited to the purposes of business planning because it emphasizes both profitability and timing. However, it is weak for controlling the progress of a development project, because its end point is far in the future, is not tangible, and is not entirely under the control of the team. Nor is it nearly as compelling for deadline purposes as is, for instance, having the product in stores by Christmas.

Although it is perfectly acceptable for the development team to do some stretching to reach the end point on schedule, this must be kept within reason and the target must be correct. One client, which manufactures products in a range of sizes, specified that one further size of the new product line should start into production each week. Management knew that this was an extremely aggressive goal, but the team did start producing one new size every week as specified. The problem was that none of these products passed inspection and it took two years of manufacturing debugging, including using lots of statistical process control, to arrive at a reliable production process for the product line.

The teams that work toward such short-sighted goals are not trying to cheat. These teams comprise functional specialists who may often view their jobs too narrowly. As in the example of the pile of parts on the assembly floor, it is management's responsibility to help them broaden their view of the project.

Once the end point of a project has been established, the planners must work through the development period and establish frequent, meaningful, measurable milestones along the way. These should be fairly frequent, because the near-term goal always seems more concrete and instills more urgency than the remote one. In a fast development project there is less time for corrections, so it is more important to have frequent goals to catch problems before they can have serious impact on the schedule. The next criterion, meaningfulness, is important because only if a goal is clearly related to an essential activity in the project will it be taken seriously.

Measurability is perhaps the most difficult feature to provide in a goal for a development project. Because innovation is a journey into the unknown, it is often difficult to know how much work lies ahead or how far one has proceeded toward the destination. The process of prototype debugging provides a good example of how difficult it can be to measure progress. When an engineer is working on a bug in a prototype it is often impossible to know whether it will be fixed in ten minutes or ten days. Then when it is fixed it may just uncover a new layer of bugs, or the fix may introduce problems of its own. It is usually impossible to know whether one has found and corrected all the bugs. *The Soul of a New Machine* provides a vivid description of the frustrations of debugging, the immense effort that can go into this activity, and the uncertainty of not knowing when it will terminate.

One way of dealing with the measurability problem is to be working backward constantly from the end point. A common situation is, for example, that the debugging is going slower than planned but the product documentation work is ahead of schedule. Only by working backward can one tell if this is really a valid trade-off in meeting the end date. If the debugging is behind schedule and that is what determines the critical path, it needs more resources, regardless of the marvelous progress being made on the documentation.

The project's intermediate goals must also be individuals' goals. The team leader should develop goals jointly with individuals to ensure that they are realistic and establish ownership. This can be a difficult step requiring a considerable amount of negotiation to arrive at an acceptable solution. It is nevertheless an essential step in an accelerated project, because until individuals sign up for the plan there is no real plan. We have seen many plans drawn up by marketing to satisfy sales deadlines, but because others did not help create the plan and thus buy in to it, the desired schedule was never maintained.

Many computer programs called project management software are available to help with the planning effort. Although most of these programs are powerful and loaded with features, they have only one that is of much value for fast development projects: being able to create a picture of the schedule, either in bar graph (Gantt chart) or network (PERT or CPM) form. The best way to use these programs is to produce a giant project schedule chart and post it on the wall of the team area for all to see. (A CAD system may be better than project management software for making these giant charts.) If the schedule is important, it deserves to be displayed prominently to signify its significance. This schedule will also be handy for the team to use in discussing progress and pitfalls, and it will be an aid when management visits to see how things are going.

Once the schedule goes up on the wall, it should stay there, in contrast with conventional uses of project management software. Normally the latter provides many kinds of reports that are to be generated on a regular basis as the schedule is updated. These kinds of reports are geared mainly to meeting government contract reporting requirements for ponderous projects and are thus inappropriate for fast development projects. Once the schedule has been created, the team leader should be spending his or her time "chatting" with team members as described earlier to control progress, not be at the computer terminal updating schedules or creating schedule reports. To track progress and problems, use colored markers or pins on the wall chart.

Formal Project Reviews

Formal reviews are a traditional means of monitoring progress on development projects. They are often called design reviews to focus them more narrowly on the technical aspects of a project. These reviews often come at critical financial decision points. When they provide a basis for making a decision on providing the next increment of funding, they may be called phase reviews.

Conventional reviews have several shortcomings with regard to accelerated development projects. For one thing, they come far too infrequently to be effective in managing daily progress. And depending on how the review is conducted, a considerable amount of time can be consumed in preparing for it, conducting it, and reacting to well-intentioned suggestions made during it.

Some companies refer to these events as tollgates, which brings to mind long queues of impatient drivers lined up at rush hour to pay whatever is needed to allow them to get on with their journey. Some turnpikes place tollgates sparsely, merely to collect enough revenue to keep the highway in operation, but others locate them every few miles, apparently for speed control. The job of managing product development is to structure reviews in a way that will ensure design quality and provide the information needed to make project-funding decisions without bringing traffic to a standstill.

Formal reviews have two functions that must be distinguished from each other. The design quality review is intended to ensure that the product will function as intended, that its cost is acceptable, and that it will be manufacturable. This activity tends to be heavily technically oriented, although it clearly must be market driven. The design quality review is basically a problem identification and problem-solving activity.

The other function of formal reviews is to evaluate progress and allocate resources. This kind of review has more of a business than a technical orientation, with the purpose of making decisions rather than solving problems. The decisions to be made are whether satisfactory progress is being made, whether the project is still a strong enough business proposition to warrant the continued funding of it, and what resources must be allocated to it over the next period to keep the project on schedule.

Again, these two types of reviews should be kept separate. Design reviews should involve technical peers rather than business managers or superiors. Try to use people from other parts of the organization who can readily understand the technical issues involved but who do not have a stake in the project. Seasoned veterans are usually best for spotting potential problem areas quickly. To keep the reviewer from becoming too critical, occasionally reverse roles so that each reviewer realizes that someday he or she will be the subject of the review. The reviewer's role should be more that of a coach than an adversary. Keep the review as informal as possible, with frequent one-on-one reviews being the ideal, using the "chatting" technique mentioned earlier. Informality both reduces the time needed for reviewing and sets up an atmosphere of trust in which the reviewer is more likely to make realistic suggestions. The reviewee will also be inclined to be more open to suggestions rather than just trying to defend a position.

Frequent reviews are best, partly to catch design problems before they can solidify and require extensive rework but also because in long, infrequent reviews the reviewer can easily be overloaded by being presented with too much material at once. If a report is needed on the review the reviewer can report orally to either the team leader or the team executive, whoever is appropriate.

Having frequent design reviews can help accelerate a project. Designs can be shared earlier, while they are still in a fluid state, and the reviewer can help to redirect efforts or confirm that the best path is in fact being pursued. Then the designer can avoid agonizing over design choices or polishing a result that may be discarded later.

As with design reviews, many progress reviews can also be handled informally, as for example by having the team leader meet weekly with the team executive to discuss where the project stands and what shifts in resources may be indicated. If the master schedule has been posted on the wall of the team area, these meetings are best held in that area, to give the executive a better idea of how the team is doing.

Major funding decision points require more structure, though, which should be built into the schedule so that the team's work can be planned so as to avoid delays caused by reviewing. Much of the delay associated with progress reviews arises when reviewers have not been prepared on the project's intent and features beforehand. When no context is provided in advance it takes reviewers longer to reach a decision once they receive the review material. If the reviewers' decision is likely to require additional funds, they should be ready for commitment in advance. It should be no surprise that product development costs money, even sometimes more than anticipated.

In this regard Xerox has developed an approach that works well for them. They use a combination of formal progress reviews and informal design reviews to control progress. At the outset of a project they take time to identify clearly the objectives and success criteria against which progress will be measured. Then they specify the timing and content of the formal progress reviews and define an acceptable envelope for the performance, cost, and time parameters of the final design. As long as the team can keep within this envelope the project flows smoothly, with funds being released as planned and formal reviews conducted as scheduled. But if the team cannot stay within the envelope or if marketplace conditions change, an exception review is called with management as quickly as possible to establish new parameters for the envelope. For this system to function smoothly management must allocate in advance the resources needed to complete the projects being started. This challenging topic is the subject of the next chapter.

SUGGESTED READING

Kotter, John P. 1990. What leaders really do. *Harvard Business Review* 68(3): 103–11. Conventional product development processes emphasize management systems to monitor and control progress through plans, reports, and reviews. Our alternative stresses providing the team with a vision and empowering and motivating the team to execute the vision. Kotter explains this subtle but crucial difference.

Randolph, W. Alan and Barry Z. Posner. 1988. *Effective Project Planning and Management: Getting the Job Done*. Englewood Cliffs, N.J.: Prentice-Hall.

———. 1988. What every manager needs to know about project management. *Sloan Management Review* 29(4): 69–73. Both of these pieces get right to the heart of managing projects for results with a message that directly reinforces the material in this chapter. The article nicely condenses the book above, which is just 163 pages itself.

Vincent, Geoff. 1989. *Managing New Product Development.* New York: Van Nostrand Reinhold. To assess progress one must understand the soft spots in each type of technology, to know where to probe. Chapter 7 explains nicely where to probe in electronic hardware and software technologies.

CHAPTER 11

Capacity Planning and Resource Allocation

Each chapter presents the development process in a new light, to gain perspective on how product development time is actually consumed and how to use less of it. This chapter spotlights resources—usually development labor, money, and facilities. Resources are normally managed so as to minimize overhead costs or enhance the return on assets or some other "hard" financial measure. Because these measures typically do not account for the cost of delays, they often lead management to make decisions that then unknowingly delay product development. The objective of this chapter is to expose such management practices, show how they arise, and offer techniques for overcoming them.

As indicated, this chapter rests on the premise that product development time has a definite, calculable financial value. Knowing what the development time is worth prepares us to avoid falsely economical practices and make management decisions that will give proper emphasis to development time. Chapter 2 gave the quantitative background for this chapter by demonstrating how to calculate the financial value of development time.

BUYING TIME

Although there are many opportunities to save development time outright, such as the techniques for overlapping covered in Chapter 9, there are also opportunities to exchange money for time. They can occur in any phase of the process and all the parts of an organization. They tend to be different for each company, so they provide a source of competitive advantage that cannot be readily copied by outside firms.

To take advantage of these occasions, the first requirement is to know the value of time for your new products, to be able to assess whether or not full value is being received for the money. Knowing the value of time will also help justify decisions, because some of these exchanges appear wasteful when evaluated with traditional accounting criteria.

The Travel Budget

Traveling to customers, suppliers, or trade shows is one way to get started on a design quickly. The objective is for the actual designers, that is the engineers or industrial designers, to get firsthand ideas about the new product that will allow them to get started on design concepts months before they otherwise might by using more conventional approaches. Traditionally, marketing visits customers, commissions a marketing study, then provides engineering with the product requirements or specifications. Engineering digests these, negotiates them with marketing, and starts designing.

Travel by designers to customers to get ideas about a new product is not the norm. Marketing usually does this—and needs to continue doing it. When we have surveyed engineers about their travel to customers, however, we have found that it is typically reactive, to support a sales call, conduct a field trial of a prototype, or most often to correct problems on units in the field.

Here we are suggesting making extra visits to customers by those directly involved in design activity, to get a head start on the normally slow front-end processes (see Chapter 3). This will probably cause some disruption or concern in marketing, and will cost money. It is still almost always worth many times the amount spent on it, though.

There are no doubt other unorthodox travel arrangements that can save time. Chapter 13 provides an example of a manufacturing foreman who travels abroad to inspect new machinery so he can accelerate its adoption in his factory when it is purchased. Here the time saving is in the tail end of the cycle rather than the beginning, and the trip is made by a person who might not normally be considered part of the development team.

Fast Experiments

The design process is a curious mixture of careful analysis and relatively unstructured experimentation. Corporate control systems drive us toward the former method, but the need for speed favors the latter, done quickly. The fast experimentation approach recognizes that many options have to be explored to reach an acceptable design, so it encourages building models and throwing them away as fast as possible, to work through the alternatives quickly. We encourage making models that are cheap and simple, because they are faster to build, but one can still burn up a lot of labor and materials this way.

Such models may cost hundreds or even thousands of dollars each and will all wind up in the trash as failed experiments in some sense, though they really are not, because the designer will have learned from them. The difficulty comes when those operating in a traditional accounting mode discover the models in the trash. There must be less wasteful ways to design, they reason.

Nevertheless, if development time has financial value, it is worth investing in quick failures that will move us forward rapidly. The motto should be to build a tall junk pile.

Clearly, moving ahead quickly by model building requires the capability for fast model building. In recognition of this fact Xerox split up its centralized model shop and assigned its technicians to work directly with development teams. They have thus lost the economies of scale, and their model makers may now not always be busy, but they are more responsive to market demands.

A similar logic applies to testing. Consider the strategy of one of our clients. This manufacturer of equipment serving the public safety market has to submit its new products to the Underwriters' Laboratories for a brutal $20,000 test. The conventional wisdom, which they had originally followed, suggested that they raise the odds of passing the test to be as high as possible, through exhaustive checking before going to UL. Once they recognized the value of development time, though, their strategy shifted to a two-test approach. They now get the design to UL quickly and let UL tell them about the product's shortcomings that need more development work. This company is now $20,000 poorer, but months ahead.

"Excess" Equipment

Normally we wait to buy a new piece of laboratory equipment until we can assure ourselves that it will have a high utilization rate to minimize the cost per hour of its use. We do not apply this logic to the jack in the trunk of a car, though, because we know that the jack will pay for itself many times over through reducing downtime if it is ever needed. The same financial logic should be applied to buying laboratory equipment. It may not be needed often, but it will pay for itself through time savings when it is needed.

Some products require building expensive handmade prototypes to check out the design, and sometimes the market. These prototypes then tend to be in high demand because there are usually more uses for them than there are prototypes. Consider a typical piece of factory automation equipment under development. At least one prototype has to be put out for a field trial. It is also helpful to have another one in the lab, so that problems discovered in the field unit can be replicated and solved simultaneously in a well-equipped lab. At the same time, marketing may want a prototype to take to a trade show. There will be delays then if a company that really needs three or four prototypes tries to make do with just one. A financial analysis involving the value of time can show what is the right balance to seek.

Farm It Out or Add Capability

Model making, testing, and even design can be done by outsiders, of course for a fee. The "outsiders" can either work at a remote site or be co-located with the design team. The choice will depend on how highly the external activity must be integrated with the internal work. If it is possible to partition the work to minimize the interface complications, as described in Chapter 6, this work can be farmed out, although it is still necessary to monitor progress closely. In either case, the justification for going outside rests on the cost of doing so and the amount of time thus saved, which can also be translated into cost terms.

Sometimes it is best to bring outsiders inside—permanently. It can actually make sense to maintain a service that is not cost effective, if it saves sufficient time. When IBM developed the PC in record time, it found that it needed manuals for the new computer faster than they could be obtained by coordinating the process through the corporation's centralized publications facility. So they set up their own publishing operation, which was more responsive, though probably more expensive.

The Tooling Game

For many products the acquisition of tooling, dies, and fixtures is a long lead-time item on the critical path. This complication certainly applies to injection-molded plastics, investment castings, forgings, and stampings. In many cases it will not be clear whether the part will be satisfactory until one produced from production tooling can be tested. If it fails—and remember that this is part of the design learning process, not a mistake—the tooling rework cycle can be both time consuming

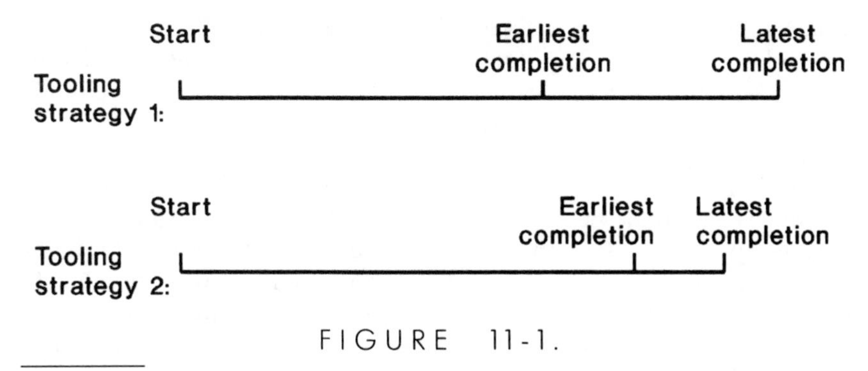

FIGURE 11-1.

Some tooling strategies lend themselves to the promise of a cycle that is short but subject to late surprises, whereas others are nominally longer but more certain.

and unpredictable. Consequently, there are two time issues associated with tooling: the amount of time needed to produce the original tooling, and the uncertainty introduced by last-minute surprises when tooled parts do not perform as expected.

Companies have developed many strategies to cope with these issues. One is to ease into initial production with partially tooled products. Even though the initial part or assembly cost may be high, some tooling time will have been cut out of the schedule. Another is to invest in tooling speculatively. In this approach commitments are made early to tooling designs that may not be perfect, to start the long lead time items, and the company is willing to invest more later for tooling rework if its early assumptions prove wrong. A third approach is so-called soft tooling, in which early production uses quickly fabricated, easily modified tooling made from relatively soft materials. Then the company invests in a second, durable, set of tools after going into production.

In each case the company ultimately spends more, but it gets to market sooner. Each situation is different, so a customized analysis is needed that not only depends on the value of time to market but also on the nature of the production processes used, the product's dependence on tooled parts, and the relative importance of having a short development cycle as opposed to a schedule free of late surprises (see Fig. 11-1).

PROJECT OVERLOAD

As consultants we visit many companies to discuss the issues plaguing their product development process. Without a doubt the issue that comes up most often is project overload. Workers typically believe they are working on so many product development projects that they cannot do justice to any of them. Because this problem is so commonly experienced, we have spent some time trying to understand and measure it, and develop solutions.

Having an excessively long list of development projects is a symptom of the fact that the financial value of development time has not become apparent yet to the project planners. Because development time is valuable, putting a project in queue to await development is financially shortsighted. This is one more example of the difference in the behavior of companies that know quantitatively what delays cost them and those that don't.

Some Barometers

The problem of having heavy workloads is so common as to be a fact of life in business. Most people would probably agree that they are trying to do too much at work, and could probably say the same for their personal lives. What is a reasonable limit, and when do we know we have undertaken too much?

These questions are answerable in terms of productivity. If we are carrying only half a load, adding more adds to our productivity. Even if we have a full load of tasks, then, it may be productive to add a few more, because some of the ones we have may not work out or we may have to put them aside for a while to wait for more information. Past this point it is not so clear what happens. What usually happens is that people are working at capacity and the extra work simply adds to the time they have to spend on overhead time, diminishing their productive working time. Even when team members attempt to spend no time on the extra work, this work may still invite interruptions from outsiders interested in it.

These issues of personal overload carry over to the area of project list overload. When do we know that we have a longer list of development projects than we can execute productively? There are three indicators.

The first is to look at individuals' workloads. Are they trying to work simultaneously on many different tasks, to keep many masters happy? How long does it take someone to complete a given piece of work? Is the common response "I can't get to it for a few days, because I have thus-and-so to do"? If those few days are on the critical path of a development project, this response has just delayed the product introduction date by that time. If this happens one hundred times in the course of a project, a delay of a few hundred days has crept in.

This problem reached even more serious proportions at one company with which we are familiar. This firm has been subject to two

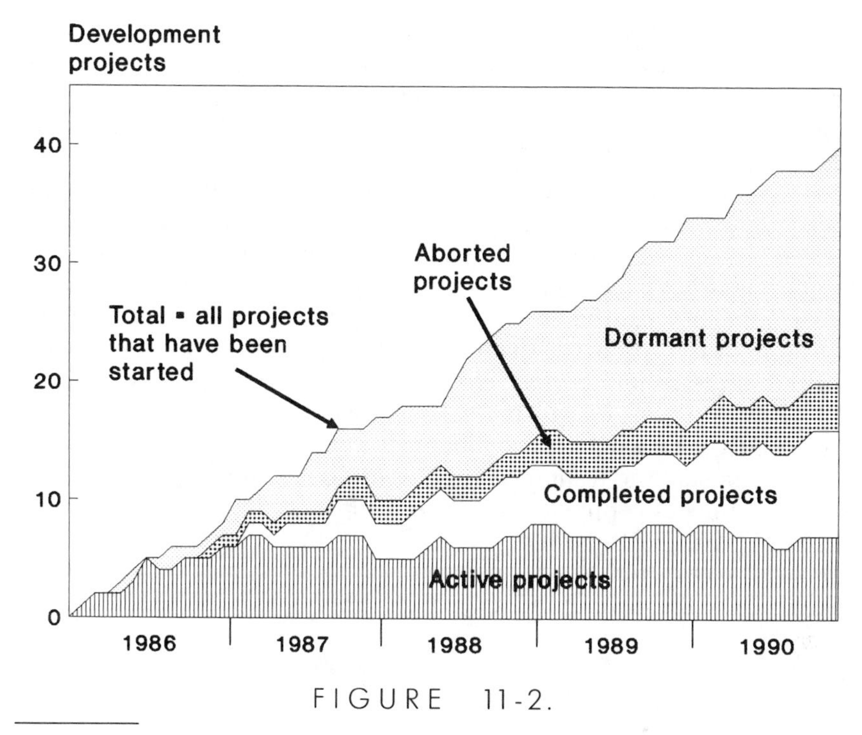

FIGURE 11-2.

Certain trends in a project list over time, particularly in the growth of dormant projects, indicate overload conditions.

buyouts in the past few years and put a great deal of pressure on those maintaining its product line to maintain its vitality, image, and profitability, as well as to merge the old product line into that of the new corporate family. There was clearly far more product development work needed than could reasonably be executed. Consequently, an understanding had evolved within the company that when someone came to you with a new task you were to accept it and say you would do the best you could to complete it. There was a gentleman's agreement that no one would be held to a schedule because all the schedules were interacting and unrealistic. People therefore worked on a best-effort basis rather than a complete-on-schedule basis. Schedules became a joke. When the end of the workday came, people put their work down and went home because an hour or two of extra effort would have made little difference. In effect, overload had reached the point of counterproductivity.

The second indicator of an overload is to look at the project list over time. At what rate are projects being added to it, and at what rate are they leaving the list? Is the list growing, or stable? Are there more dormant projects than active ones? Figure 11-2 is a project list diagram illustrating some of these issues. The top curve in the diagram represents cumulatively all the projects given approval to start, with the area under this line accounting for the outcome of each project. Some have been officially aborted, some completed (resulting in new products), some are still active, and some have become dormant (with, for example, no effort having been expended on them for at least three months). This diagram is fairly typical. Few projects are actually being aborted, the completion list is growing steadily, and the active list is fairly constant because it depends on product development capacity. However, the dormant list is growing steadily because projects are being added faster than they can be completed. The projects on the dormant list consume overhead energy that could have been put to work on active projects, so it is counterproductive.

The third earmark of project overload is delay in starting new projects. How long does it take once a project is approved for development to marshal the resources to put a full development team at work on it (a couple of people just piddling with it does not count)? In most companies people are so heavily committed that it takes months to get up to full team strength on a new project, months that add directly to the length of the development cycle.

The Project Load Myth

As suggested, simply having a large list of dormant projects drains energy from the active ones, but the impact of trying to work on too many projects is far more dysfunctional than this and can be measured financially.

To demonstrate quantitatively what the impact of overload is, let us extend the financial modeling approach covered in Chapter 2. (A review or browse through Chapter 2 at this point might prove useful.) Chapter 2 analyzed a single project in isolation, but here we consider a portfolio of such projects that comprises all the projects initiated in a particular year.

Assume that while we are working on such a portfolio a new, urgent, project comes along. It is 10 percent of the size of the portfolio. It is impossible to displace any of the critical projects in the portfolio, so the 100 percent portfolio now becomes a 110 percent portfolio. Also, the product development capacity was fully utilized already by the 100 percent portfolio. Undertaking a 110 percent portfolio then implies inescapably that introducing the whole portfolio will be delayed by 10 percent.

The underlying assumption is that a company develops new products ultimately to make money for the company. Thus, the company decides to go from a 100 percent portfolio to a 110 percent one because its managers believe there will be a positive profit impact for the company, although they accept that it will probably not be the full

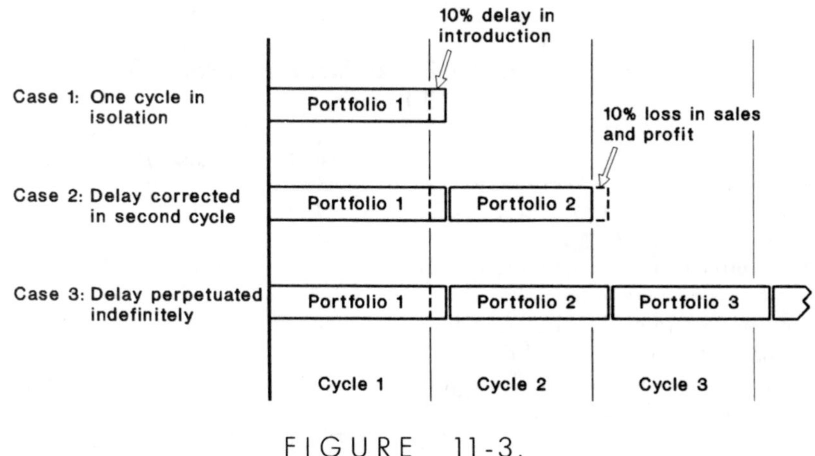

FIGURE 11-3.

Three different outcomes or cases when an overloaded project portfolio (#1) is increased by 10 percent.

10 percent. Our consulting experience suggests that getting a positive profit impact out of an overload situation is usually a myth. Some calculations will show why.

Consider three different cases to illustrate how the impact plays itself out, as illustrated in Figure 11-3. In the first case, just one development cycle is considered, in isolation. Ignore for now the fact that if this portfolio is late in completion the following portfolio will have to start late.

Cases 2 and 3 are two different ways of dealing with the impact on the following portfolios. Case 2 is a highly disciplined approach in which the following portfolio is cut back to 90 percent to correct for the schedule overrun and return to the normal cycle. In Case 3 no correction is made, so that all successive portfolios continue to be delayed.

Assume that the development cycle is twenty-four months, as in Chapter 2, so that a 10 percent delay means 2.4 months. The other input values also come from the results in Chapter 2. The cost of delay is $470,000 per month in pretax profit for a project with a cumulative pretax profit of $10,400,000. Consequently, 4.5 percent of the profit is lost per month of delay, or 10.8 percent for 2.4 months. Table 11-1 illustrates the net impact on profitability of this delay.

Case 1 provides the basic message. By squeezing in one more project, presumably to improve the company's profit picture, and ignoring its effect on future portfolios, the net profit from this portfolio of products is actually decreased.

As one might expect, the situation only gets worse in cases 2 and

TABLE 11-1.

The Impact on Profitability of Increasing Portfolio 1 by 10 Percent.

	Case 1	Case 2	Case 3
Profit impact in cycle 1 due to a 10% larger portfolio	+10%	+10%	+10%
Profit impact in cycle 1 due to a 10% extension in cycle length	−10.8%	−10.8%	−10.8%
Profit impact in cycle 2 due to a 10% smaller following portfolio	—	−10%	—
Profit impact in cycles 2,3,4, . . . due to a perpetual 2.4 mo. delay	—	—	− 41.3%
Net profit impact	−0.8%	−10.8%	− 42.1%

*Net present value using 15% annual interest rate

3, which are probably more realistic. Case 2 suffers a 10 percent profit decrease in the following portfolio, which is charged to the initial portfolio, since it was the greed in planning the initial portfolio that precipitated the whole problem.

Case 3 is more complex to calculate because of the need to account for a continuing string of delayed projects. Chapter 2 avoided using net present value analysis because all its events occurred over a relatively short interval. Here, though, the profit loss comes from having an indefinitely long sequence of delays, so net present value analysis is appropriate here to account properly for the fact that in the future funds will have less value than those available today. If each follow-on portfolio is considered to be delayed 2.4 months and a 15 percent annual interest rate (a discount rate of 13 percent per year or 24.4 percent per two-year development cycle) is applied, there is a 33.6 percent profit loss in the following portfolios, all chargeable to the decision made in Portfolio 1 to add just one more project.

There are more sophisticated ways to analyze this problem, but its conclusion rarely changes: extra projects rarely result in extra profits when an organization is already running at full capacity. Although this example is not atypical, neither is it likely to be precisely correct for another company's products. The values resulting from such analyses can differ greatly in various industries and companies, or even for particular products within the same company. Doing the calculations for your own products will provide more accurate results and give you a feel for how the numbers fit together, thereby increasing your understanding of the underlying financial mechanisms.

The Mushroom Effect

Because the project overload trap is a dangerous one we need to understand how it works so we can spot it and avoid it. This trap exists because the tail end of the product development pipeline has more serious constrictions in it than the front end. It is relatively easy at the beginning for marketing to do its market research, make a return-on-investment analysis, write up a product specification, and issue a request for a new product. Then the design, development, and engineering processes require more detailed work and thus more labor to turn that product request into a set of part and assembly drawings for manufacturing. There the project mushrooms even more because a great deal of work is needed to turn each part drawing into a stable manufacturing process. Figure 3-3 illustrated this by showing a typical project-spending diagram. Although some of those funds must go toward material, such as tooling or equipment, the spending also includes a growing labor portion, representing the ever-larger number of people involved in a project as it progresses.

Manufacturing people tell us in our consulting work that they simply have to react to the drawings given them by engineering and do the best they can, because engineering is driving the process. Similar discussions with engineering reveal, however, that they really are not in control but are just reacting to instructions sent down from marketing. Of course, if the joint specification writing techniques suggested in Chapter 5 or the early manufacturing involvement approaches covered in Chapter 13 are used, the problem diminishes, but the seeds for a mushrooming project are still there.

It is important for marketing to understand that they are in fact the ones driving the process and that they can easily push it into overload. One of our clients provides a graphic example of how marketing can control a situation. This client had a serious overload problem, so we did a financial analysis similar to that shown in Table 11-1 to illustrate that carrying so many projects was actually reducing their profit. We then went over these results with the marketing director, who was just then in the process of preparing the next year's project list for the executive staff to review the following day. As we talked, he came to recognize that it was his own project list that was driving the whole organization into overload. When he made his presentation to the staff the next day, he showed them the list of twenty-three projects he had planned for the year, one similar to prior annual project lists. Then he said that he appreciated that they were already working on too many projects and proposed to cut his list to just six projects. Furthermore, if engineering and manufacturing still had reservations about being able to complete these six projects, he said he was willing to cut his

list to only three projects—an 87 percent reduction—in return for their assurances that the projects on the new trimmed list would be completed on schedule.

Marketing is under its own pressures. It must be responsive to the sales force. Each salesperson would like to have a new product that will relate to the order he or she lost the day before. Marketing must sift through all this conflicting information and prepare a product plan. The department is usually evaluated—and this is the key point—on the comprehensiveness of the plan they present. It is not very satisfying, nor does it seem to reflect well on the product planning group, if they present a plan for only three new products when the sales force genuinely needs twenty-three new ones. If the company has a track record of introducing only three new products per year, however, then this is its capacity and twenty-three products is merely a dream.

We need to be careful about evaluating the work of the product planning group. And they in turn will have to work harder and subject themselves to more criticism as they shorten their list.

Avoiding Overload

The basic means of avoiding overload is to start fewer projects but get them done, then start a few more. To put this basic message across, here is a simple, idealized model. Suppose that a company has a fixed amount of development resources, which we will take to be the R&D head count, as represented on the vertical axis in Figure 11-4. There are two projects on this project list, with time portrayed on the horizontal axis. There are two options, as illustrated. The more commonly chosen one is to divide the resources in half and work on both projects simultaneously. The other option is to put all the resources into one project, get it done, then put all the resources into the other project.

The illustration shows the second option ending at the same time as the first one. In reality, the second option may take longer, which may occur if there are fixed time constraints such as tooling lead times driving the development cycle. If the limiting constraint is the amount of effort (labor hours) that can be put into the project, the second option will end at the same time as the first because the labor hours (the areas within the rectangles) are exactly the same in the two options. Notice that when working in an overload mode, the limitation is labor hours, and the situation illustrated, with both options ending at the same time, is accurate.

Operating according to Option 2 enjoys several advantages. One is that the first product gets to market in half the time, as indicated. Another is that it provides more flexibility because no commitments need be made to Project 2 until Project 1 is finished. At this point it

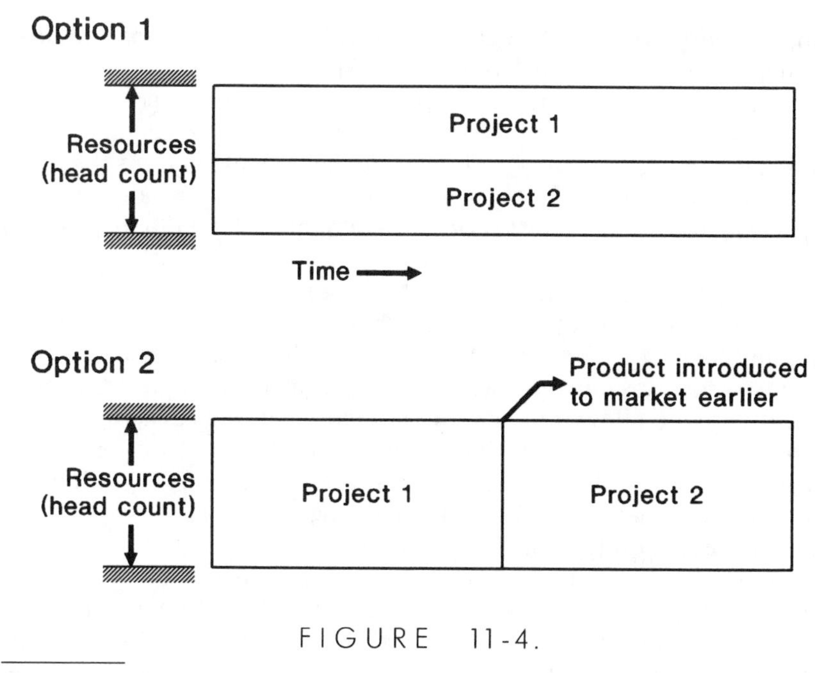

FIGURE 11-4.

Simplified model contrasting simultaneous versus sequential effort on two projects.

might seem best instead to do Project 3, or do Project 2 in another way. A third advantage is that if you do decide to do Project 2, you will be doing it with more current market and technical data than before. The final advantage is that of cycling people through projects faster so that they learn more and keep their skills sharpened.

In advocating Option 2 we may seem to be contradicting the advice in Chapter 8 to keep the development team small, to facilitate communication. Certainly, carrying this concept to extremes and putting all the staff on one mammoth project would be inefficient. What tends to be the case in industry, however, is that even individuals and small groups tend to operate according to Option 1 rather than 2. These people are fragmented, splitting their time, and when they are working on two tasks simultaneously it will take them roughly twice as long to complete their tasks as if they had worked on them individually.

Unanticipated Projects

Another reason for starting fewer projects is that it is counterproductive to run a product development operation at 100 percent of its capacity with known projects, because some slack must be provided for handling unanticipated projects. Such projects are not unwanted distractions or signs of an undisciplined organization, merely an indication that the firm is responsive to market demands. Chapter 3 discussed the type of annual planning cycle that completely fills the product development pipeline for years in advance, thus virtually assuring that every project will start serious development long after its market need is identified.

Manufacturing planners do not plan plant capacity to run at 100 percent of current demand because they know that the demand will shift and they will need some flexibility to satisfy changing market conditions. When it comes to product development, however, the same executives who allow for a margin on manufacturing capacity seem to be fearful of having engineering resources sit idle. They neglect to factor in the cost of delaying the market introduction of new products in the same way they allow for manufacturing overcapacity so that they can ship orders on time as the demand changes.

Operating at full capacity on planned development projects causes another type of dysfunctional operation. Unanticipated projects will always come along and demand an immediate response, and current projects will sometimes require more effort than planned. Then the organization will have to shift people and priorities, which always destroys productivity and commitment. Although a certain amount of shifting is inevitable, by not allowing some slack the development efforts will be in constant flux.

Project Priorities

Shifts in development priorities are indicative of project overload. We have often found that priorities among development projects are unstated, uncommunicated, or constantly churning. What happens when management does not send clear, steady signals about its priorities is that individuals then establish their own, based on various individual objectives. Then people are working at cross-purposes.

We are continually amazed at the number of companies that have no priority list at all for their development programs. A good way to instill some discipline in setting priorities is simply to post the priority list for all to see. Clearly, there are some proprietary information issues in posting lists of product development projects, but these usually amount to smoke screens for not committing to ongoing priorities. Projects can be given code names if necessary.

WORK IN PROCESS

Chapter 9 mentioned that just-in-time (JIT) manufacturing techniques provide several useful analogies that carry over to the product development process. JIT puts great emphasis on shortening manufacturing cycles, which results in having reduced inventory requirements, especially those for work-in-process (WIP) inventory. WIP is the stacks of partially finished parts to be seen when visiting a factory floor. The product development process has its own WIP too, but it is not quite as obvious. WIP is the paper in everyone's in basket, the backlog schedule in the testing lab, the model shop, and tool room, as well as the list of unstarted projects in the annual plan.

WIP exists because of insufficient capacity or resources at the next step to process it. In manufacturing it is a sign that a plant is operating under a philosophy that tries to keep all the machinery busy all the time to maximize the return on assets, rather than to ship quality products on a responsive basis. JIT experts say that in an effort to fine tune the profitability equation we have gotten our manufacturing priorities backward.

The same can be said of the product development process. In an effort to maximize the return on the product development assets—equipment and people—we have created queues of WIP everywhere in the process so that assets will always be productively engaged. In doing so we have lost sight of the fact that time matters and backlogs cost money. New products sit as WIP while the market window, which provides the value, is closing.

JIT can teach us more about WIP as it applies to new products. One of the most important tools to come out of JIT is the technique of value-added analysis, in which a single part is tracked through its manufacturing process step by step to see how much time the part actually spends gaining value. Some of the non-value-adding steps are transporting, stacking, cleaning, and inspecting. Typically, it is found that 80 percent or more of the manufacturing steps do not add value. More apropos to product development is that manufacturing's value-added analyses typically find that a part spends less than 0.5 percent of its time gaining value. Starting with such analyses, companies have been able to make dramatic reductions in manufacturing's lead times. Similar analyses can be applied to the product development process to show how much time designs spend sitting around not gaining value.

Analyses can show us where there are opportunities to save time and even how big they are. The more difficult part is in making the changes indicated. In the case of reducing WIP, the changes are difficult to make, whether they apply to manufacturing or product development, because they require changing long-held notions of proper management practice.

For new product development the standard management practice is to make sure that all employees are always fully engaged, even though sometimes this is not the most effective way to get new products to market. For example, one client developed complex machines having hundreds of new parts. There was a point in its development when the engineer had arrived at producing the design for the machine, but detail drawings were needed for hundreds of parts, which overloaded the small detail design staff. The company thus hired outside contract detailers, which did not work out very well because the contract workers were not experienced with the particular CAD system the company used, nor did they know its drafting standards. The engineers therefore spent an inordinate amount of time during a critical phase of the project in supervising outside workers.

A more effective solution, though it violates normal management principles, would be to have more detailers on staff permanently. Even if they only sat around and drank coffee for one week out of the month, when needed they would be so much more effective than outsiders unfamiliar with the system that it would more than offset their idle time.

Manufacturing managers have been learning nonintuitive approaches like this through JIT. Product development managers need to develop a similar attitude toward questioning the accepted management processes in their own departments.

THE COST OF SPEED

Chapter 1 briefly addressed the fear that many managers have about speeding up the development process, that it will drive up R&D expenses. This fear is well founded. The paper by Graves listed at the end of this chapter argues that for each percent of time cut from the development cycle the R&D expense will typically rise by 1 to 2 percent. Moreover, we have seen earlier in this chapter how one can "buy time" by making time–money trade-offs wisely.

There are two other points to consider, however. One is that a firm's management must understand why they are developing new products before they can properly decide whether to minimize expenses or time. In the true spirit of making time–money trade-offs, it may indeed be most profitable to pay an additional 1 to 2 percent in expenses to save 1 percent in time. Managers who focus only on R&D expenses may be focusing too narrowly. One can also concentrate too narrowly, however, by fixing only on time to market. If a company's product line is broadly in need of upgrading but the marketplace is not demanding new products quickly, the proper objective might then be to produce the greatest amount of product development for the available R&D funds. But without a well-articulated development strategy and some financial analysis to support the decision making such discussions can go on endlessly.

The other point is an even more significant one. By going beyond trade-offs and fundamentally improving the management of the product development process, a company can obtain savings in both time and costs. Chapter 1 mentioned several companies that have achieved both goals simultaneously. Closer, more informal communication, lower-level decision making, and the improved commitment that comes from having dedicated project participants all will act to improve the speed *and* effectiveness of the product development activities. For instance, in Figure 1-5 such improvements in effectiveness basically result from a reduction in bureaucracy.

Such improvements eventually reach their limits, however. Then, as one continues to cut time out of the development cycle the costs will rise, as was indicated in Figure 1-5. The article by Graves at the end of this chapter provides many reasons for costs to rise. In this short-cycle region we are operating in a domain of inefficiency created by excessive staffing, using redundant efforts to overcome uncertainties, and starting activities before enough information is yet available to work on them productively.

Our consulting experience suggests that most companies operate in the right-hand region of Figure 1-5 so that they have opportunities available to them to save both cost and time. Companies should be operating somewhere in the left-hand region (not at the bottom), depending on their own cost–time trade-off points.

SUGGESTED READING

Graves, Samuel B. 1989. Why costs increase when projects accelerate. *Research-Technology Management* 32(2): 16–18. Graves clearly has a different view of accelerated product development than that expressed throughout this book, but this paper bears careful reading. Only by avoiding some of the conventional management approaches that he describes can one expect to accelerate the development process without increasing its costs.

Hayes, Robert H., Steven C. Wheelwright, and Kim B. Clark. 1988. *Dynamic Manufacturing: Creating the Learning Organization*. New York: The Free Press. Chapter 10. Although this chapter covers many different subjects, it contains informative discussions on managing the "funneling" of product ideas into a few development projects and some of the issues involved in project overload.

CHAPTER 12

Managing Risk

There is a complex interaction between development cycle length and degree of risk. One can make dramatic improvements in time to market without significantly increasing risk, or can increase risk substantially while realizing only small gains in development time. The difference is in how the risk is managed. Managing risk effectively is an important part of developing products quickly.

Increased risk is a natural consequence of the steps taken to shorten the development cycle, but taking risks does not in itself save time. What happens is that we expose ourselves to certain risks to gain speed, for example, by using partial information to overlap tasks, then use specific management approaches to limit the effect of this risk taking on the development cycle.

Product development risk is an uncomfortable subject for many people, who see innovators and entrepreneurs as fearless souls willing to bet the company on an untested idea and take shortcuts to be the first to market. Tom West, in *The Soul of a New Machine*, projects some of this image. This romantic view of technical heroes does not fit in well with reality, though. Consider what Peter Drucker has to say about entrepreneurs and innovators:

> A year or two ago I attended a university symposium on entrepreneurship at which a number of psychologists spoke. Although their papers disagreed on everything else, they all talked of an

"entrepreneurial personality," which was characterized by a "propensity for risk-taking."

A well-known and successful innovator and entrepreneur who had built a process-based innovation into a substantial worldwide business in the space of twenty-five years was then asked to comment. He said: "I find myself baffled by your papers. I think I know as many successful innovators and entrepreneurs as anyone, beginning with myself. I have never come across an 'entrepreneurial personality.' The successful ones I know all have, however, one thing—and only one thing—in common: they are *not* 'risk-takers.' They try to define the risks they have to take and minimize them as much as possible. Otherwise none of us could have succeeded. As for myself, if I had wanted to be a risk-taker, I would have gone into real estate or commodity trading, or I would have become the professional painter my mother wanted me to be."

This jibes with my own experience. We too know a good many successful innovators and entrepreneurs, not one of whom has a "propensity for risk-taking." . . . The innovators we know are successful to the extent to which they define risks and confine them. (Peter F. Drucker, *Innovation and Entrepreneurship*, Harper & Row, Publishers, Inc., 1985, p. 139).

KINDS OF RISK

Managers of product development are concerned about the total cost of a new product project to their business. The most immediate, direct cost is development expense, which is especially painful and visible if a product is a failure. A less immediate cost is what economists would call the opportunity cost of putting out an inferior product. It is the shortfall between the profits that could have been achieved compared to those actually reached. The lower profits could have been caused either by decreased sales or higher costs from inefficient manufacturing, or even by product recalls or product liability problems.

In order to manage the risk of having excessive total cost, it is useful to distinguish between two types of development risk: technical risk and market risk. The dividing line between these is marked by our old friend the product specification. Technical risk is the probability of failing to meet the performance, cost, or schedule targets of the specification. It is the risk of poor technical execution, of missing the target.

Market risk is then the probability of not meeting the needs of the market, assuming that the specification has been satisfied. It is the risk of selecting the wrong target. This distinction is useful to make because it focuses equal attention on technical and marketing activities and highlights the importance of the specification, as will be seen.

Technical Risk

Organizations usually place more emphasis on technical risk than market risk, largely because technical risk is more apparent and is easier to manage. Technical risk can arise from the technology failing to perform as expected, from its cost being higher than projected, or from unanticipated side effects that arise from a particular technical approach. Companies are more accustomed to dealing with technical than market risk and have developed techniques for assessing technical risk. For instance, they often monitor product and development costs, build checklists of technical problems experienced previously as a check on a current project, or create backups for a questionable technology.

The outcome of having to resolve technical problems is usually schedule delay. If a technology does not perform as expected, putting in a little more time to fine tune it will probably overcome the difficulties. If product cost is the problem, the design can probably be cost reduced through further refinement. And if there are difficulties with unwanted side effects, a different variant of the technology will probably work, but applying it will entail backtracking.

Some companies focus on the performance and cost dimensions of technical risk and manage them heavily. They conduct many design reviews by several levels of management, generate numerous plans that must be submitted and approved before design activity can start, and require regular written progress reports. These controls can indeed catch performance and cost problems and minimize surprises, but this management overhead consumes a great deal of time, especially if high-level decision makers are involved in it.

The net effect of having an inappropriate level of risk management is that it adds time to a project, either through reacting to design problems that could have been anticipated or by putting so many risk filters in place that the project's flow slows to a trickle. The effect of these two approaches on the length of the development cycle is different, however. In the undermanaged case the basic cycle can be short, but it will extend an unpredictable amount because of unanticipated problems that occur. Figure 12-1 shows this effect. In an overmanaged

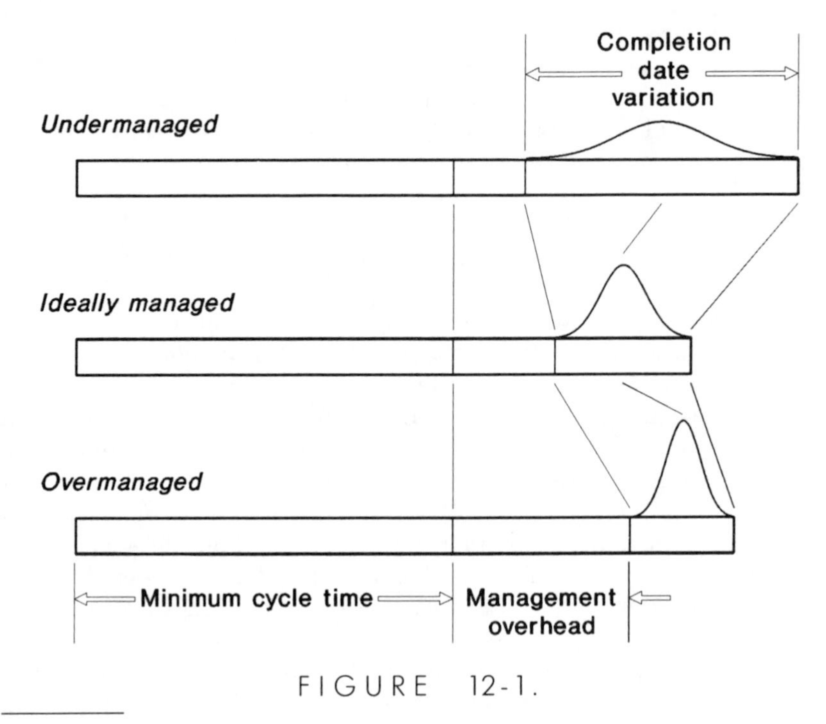

FIGURE 12-1.

By putting more time into managing risk in a project the variation in completion dates shrinks, but management overhead extends the cycle's length.

situation the cycle bears an extra burden of management time, but the variation at the end of the cycle is smaller, because the management process filters out some of the problems.

Techniques for avoiding either of these options follow shortly, but there is more to be learned from these two options, because there is a preference to be considered between them. This preference depends on the timing demands of the business. In some markets the objective is just to get the product to market as quickly as possible. This emphasis is typical of technology-driven markets such as computers. In *The Soul of a New Machine* the Eagle team was in a race with a North Carolina group and competitors like Digital; every day counted.

Other markets present specific market windows. Snow throwers have to be on the market by late fall. Wristwatches have two windows, December for Christmas and Hanukkah, and June for Mother's and Father's day and graduation. In this case the minimum cycle length is not crucial, but variations in cycle length will be costly, especially when advance commitments have been made with distributors, dealers, and advertisers. In selecting ways of managing development risk, keep in mind just what the goal is in trying to achieve fast development. In some cases it may be advantageous to add a few extra checks to the process, to minimize tail-end variations.

Market Risk

Although market risk often receives less attention than technical risk, it is no less important. Market risk can result from having done inadequate market research, from following a specification that fails to define adequately what is needed, or from unclear customer desires, changes in customer requirements, or the introduction of competing products.

Compared with how they handle technical risk, most companies do a poor job of managing market risk. This is largely because they consider it less objective and quantifiable than technical risk. Yet market risk is a more important cause of product failure. Typically, companies resolve technical problems—or agree to ignore them—before a product is introduced. In contrast, market problems usually show up only after a product has been shipped.

Consider how your own company deals with market risk relative to technical risk. How vigorously, and when, does it react to a new product that is, say, 20 percent high in its unit manufacturing cost, compared with one that is 20 percent below target in its first-year sales volume?

Market risk issues are common. Chapter 5 presented three examples of weak product specifications, two of which illustrated market risk. In the electronic instrument example the technology was available, but no one could specify what the customer wanted in terms of the instrument's accuracy. In the food-processing machine example the product was designed to the specification, but then it was discovered that the specification did not reflect the food-processing quality that customers would normally expect. In both cases the weakness was in not defining a product to satisfy customers, as opposed to an inability to meet the product definition.

The most obvious way of resolving market risk is to increase involvement with the customer, either through formal means like market research or more direct means, such as establishing routine designer contact with customers or actually having a leading-edge customer on the development team. Although market research will always be needed, let us emphasize the direct approaches because they are underused and provide a quick means for the design engineer to remain surefooted as the terrain shifts in the marketplace. Market research often does not provide specific answers fast enough, particularly in the midst of the development process, when the inevitable design trade-off decisions arise.

In addition to having frequent customer contact while developing a product, get back to the customer often with a product. Keeping development cycles short allows market research to be done in the best possible way—in the marketplace and with your own products. The technique is to get a product out, see where it falls short of the customer's expectations, make a few adjustments, and quickly introduce an improved product. This is the routine of incremental innovation covered in depth in Chapter 4.

There is a direct relationship between market risk and the length of the development cycle. The longer the cycle, the greater the opportunity is for the market to drift away from the product specification, the greater is the likelihood that the developer's thinking will drift out of touch with the market, and the greater is the chance that a competitive product introduction will invalidate the forecast assumptions.

One last suggestion for controlling market risk is to stay flexible on unresolved issues. When it is impossible to determine the correct path, build in flexibility. In the first IBM PCs, for example, the designers were not sure just how the user would store data, so IBM provided both a cassette tape port and a floppy disk option. The cassette port was never really used, but having it allowed IBM to proceed with development while the magnetic-storage alternatives sorted themselves out.

Black & Decker provides an example of combining flexibility with an aggressive application of overlapping activities (see Chapter 9 on overlapping). After developing several cordless drills the company decided to add a cordless screwdriver to their line. So they developed a cordless screwdriver with a pistol grip, which was introduced in June 1985. It was a powerful unit with a comfortable grip that could be used for light drilling as well as for driving screws. The only problem was that it failed to meet sales expectations. Black & Decker later found out that customers did not consider it a screwdriver simply because it did not look like one. A competitor's product, shaped like a conventional screwdriver, was taking away their sales. As a result, B & D initiated an accelerated eleven-month development project to provide a new cordless screwdriver for Christmas 1987.

The technical team first went to work on the functional issues. Meanwhile, marketing could not decide on the handle shape, which was a long-lead-time part, because of the necessary mold-fabrication time. They conducted some focus groups on handle size and shape, knowing this to be a critical issue. However, by the time marketing had to commit to a handle shape, they were still unable to decide between two options: a round shape and a somewhat fatter handle with a squarish cross-section. The square handle had the advantage of being able to hold three rather than two cells for greater power.

Because of the open issue on handle shape, Black & Decker decided to design tooling that could accommodate both shapes, while they did more market research. The product design was then purposely kept modular, with common interface dimensions (see Chapter 6 on product architecture) so that different handle shapes could be used interchangeably. Similarly, the tooling of the handle molds was designed so that molds for either shape could be used in a common mold frame. The product went into production at full rate in August, having square handles, to build a supply for Christmas. But when the final market research showed that the smaller, round handle was preferred, production switched over to it in October. (The square-handled tooling found an application later in a more powerful step-up version with three cells.)

Without careful planning and a sound understanding of the critical parameters of a product, the outcome can be much less favorable. For instance, one company that builds automated assembly equipment got itself into a trap with product flexibility problems. They were developing a machine to assemble a new generation of components, which were so new that no industry standard had yet been established on their configuration or size. The engineers had found a clever way to transport these components quickly and cheaply. Their elegant technical solution carried high market risk, though, because the transport scheme was sensitive to variations in component size and shape, which were precisely the issues that remained unresolved in the marketplace. Consequently, the company found itself being unable to move until the market stabilized because they had not maintained their flexibility on a key parameter that was still in flux.

Maintaining flexibility and innovating incrementally are separate techniques for minimizing market risk, but they can work well together. It is possible to use unresolved market issues to guide the path of incremental innovation, as Hewlett-Packard is doing now with handheld calculators. They initiated the concept of reverse Polish notation (RPN) in calculators, while competitors were using an algebraic system of data entry. Current HP models are available in either system, however, and their sales figures will tell them which way to go in the future.

Market and technical risk can also interact to amplify each other. In concentrating on technical risk many companies overemphasize testing or review systems to cope with this risk. In the process they stretch their development cycle, which then increases market risk. Be-

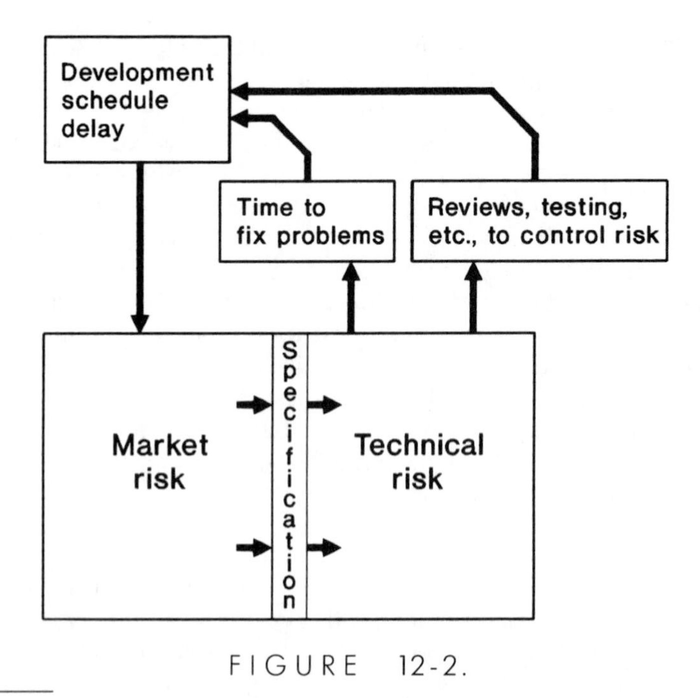

FIGURE 12-2.

High technical risk can delay a project, which in turn increases its market risk.

cause market risk is less apparent than technical, this is a subtle trap. The temptation is to resolve market risk by specifying more advanced features. Figure 12-2 summarizes the relationships discussed here between market risk, technical risk, and the length of the development cycle.

TECHNIQUES FOR CONTROLLING RISK

Many of the basic ideas for controlling risk are covered elsewhere in this book, particularly in product-related chapters 4 (on incremental innovation), 5 (on product specifications), and 6 (on product architecture). For example, Chapter 4 describes the destructive effects of the  megaproject. One client who seemed to be locked into such projects once explained that because they are so important to the company, megaprojects have to have top management involved in every decision, to minimize technical risk. "We have to support our people; it is company culture not to let them fail," he said. This interlocking of megaprojects with a fear of failure leads to high market risk. Such products are often technical successes but market failures.

The principle of keeping it simple applies from the overall product level of avoiding the megaproject right down to individual components. Whenever possible, employ standard components and proven parts—they may not be as economical to make or as technically elegant as a new design, but they impose far less technical risk.

Also, keep as much risk as possible outside the development process, with a strong program of ongoing, product-oriented technical and market research, covered in more detail in Chapter 3. It seems too elementary even to mention that companies need to do their homework before starting in on a development project, but we are continually surprised at how often this issue arises. Marketing people often wear many different hats, and market research that is in anticipation of product needs and unrelated to a specific product seems to receive a low priority. Another hindrance is that in some organizations the marketing people tend to come from the sales organization and do not have the analytical skills needed to do market research. The complex quantitative and qualitative techniques taught in business schools may not be needed for most market research, but knowing them can prove invaluable in some cases. When companies view marketing as a job that can be done by amateurs, they tend to get amateur marketing.

On the technical side, companies often do not have a technology development plan to prepare the underlying technologies for possible products. Even if they do have such a plan, it is frequently out of date or being poorly communicated to the people working on lab bench who are developing the technologies. This results in a mismatch between the technologies currently being developed and the technology needs of upcoming products.

Provide Focus to Management's Attention

Risk needs to be managed actively, both in advance of and throughout the development process. Areas of high risk must be identified ahead of time, localized to the greatest degree possible (see Chapter 6), allocated the best technical or marketing talent, have their problems resolved as quickly as possible, and be provided with an up-to-date backup plan in case something cannot be resolved.

Too many companies let risk management be a reactive process, which is simply too slow. On one assignment the objective was to discover the differences between certain slow projects and some fast ones. We found that in the slow projects glitches would occur and then the team would react to them without prior preparation, but the fast projects were managed more proactively, so that the team knew how to avoid glitches or what to do if they appeared. They tried to keep the inevitable product development glitches from becoming surprises. In another assignment, where some projects were getting bogged down in manufacturing, the manufacturing engineer, whom everyone assumed was placing a priority on managing the risk, was actually devoting his attention to a JIT program and reacting to new product problems only when they arose. Each time one did come up it took weeks to get the project back on track, because the manufacturing engineer had devoted no time to preparing for problems.

Good risk-control systems minimize risk by keeping the high-risk areas highly visible. Such systems allow management to make major resource shifts in response to emerging problems. Because in an accelerated project there is less time available to react, and because the cost of fixing a problem is much lower if it is caught early on, having early warning systems in potential risk areas is crucial. This is where management by wandering around (MBWA) pays off (see Chapter 14).

Freedom to fail is another factor in keeping an early warning system vital. Communication on potential problem areas must be kept open, both horizontally and vertically. After all, product development is no more than a sequence of problems that arise and have to be solved, and if people are discouraged from talking about problems nothing much will happen. In particular, if the messenger is shot for bringing bad news, the early warning system will disintegrate. Shooting the messenger recently contributed to product recall that cost its manufacturer nearly a half billion dollars.

Work Concurrently on Technical and Market Risk

Both technical and market risk need attention, which should be provided concurrently. The traditional development approach is to work on them alternately. Companies frequently work on market risk first, through market studies. Then a specification is written and marketing basically waits while the technical risk is reduced through product development. When there is a prototype to work with, marketing goes out for market tests to resolve market risk. Finally the project lands in manufacturing, where technical risks in the manufacturing process are resolved. Figure 12-3 illuminates the time that is wasted in this sequential approach. Chapter 9 addresses approaches for executing sequential activities more in parallel.

Although the market and technical risks interact and should be reduced concurrently, it is usually best to resolve them independently

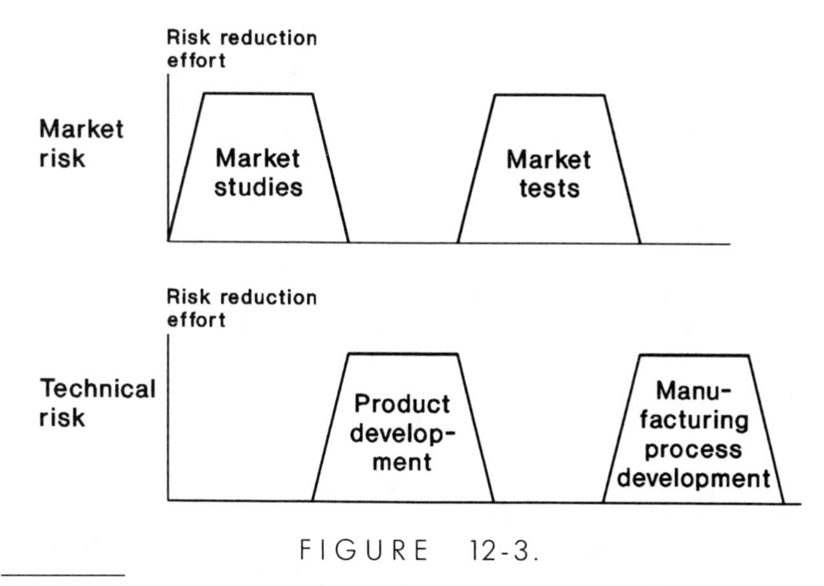

FIGURE 12-3.

Typically, companies work at reducing market and technical risk alternately, when they could be working on them simultaneously.

of each other, because acting on them separately allows for generating better information faster than by combining market and technical testing. Here the product specification provides a convenient separator to help divide risk assessment into two relatively independent parts. Then we can focus different types of management attention on each part: does the product meet the requirements of the specification, and does the specification reflect accurately what the customer wants? These two questions are often answered through testing, but it is easier—and faster—to test them separately, market tests through a non-functional styling model, technical with a crude-looking functional model. Combining the two merely muddies the management process and complicates the model. For example, in the two-track process a designer of video games would test the "play" of the games on a powerful computer simulation, but independently work out the handheld unit's shape and control feel by working with its potential users.

Maintain Backup Positions

An essential part of the initial planning for a new product is making an assessment of both its technical and its market risk in what is sometimes called a feasibility study. These identify the areas that might cause trouble, assess how likely the problems are apt to be, and estimate their consequences if they do occur. Then it is necessary to plan a backup position for each critical item. Occasionally it will not be possible to provide a backup for a contingency because there is just no other way to handle the item. If it is truly crucial to the product concept and there can be no backup for it, then put the best people on it and keep it visible.

Keep backup plans up to date. New risks normally arise as technologists get deeper into the subject and marketers learn more about customer needs. Add these new risks to the overall plan, and develop backups for them. At the same time, routinely review the plan for pruning opportunities. Because it usually requires resources to maintain active backup positions, when a risk decreases to a tolerable level cancel the effort.

Having backup plans is standard operating procedure in most development projects, especially in ones with tight schedules, but it may still be worth examining a few examples to illustrate how backups work. Dynapert is a manufacturer of equipment to mount electronic components on printed circuitboards. A few years ago they started developing a new family of what are known as through-hole machines, in which the electronic components pass through holes drilled in the board. They wanted to develop their first product in this new line in just fifteen months, half their normal development time. Moreover, they had ambitious technical plans, including one for reducing the noise of the machine dramatically to comply with their Swedish customers' new noise standard. The major noisemaker was the pneumatic head that drove the electronic component's leads through the holes in the board. Dynapert's plan was to replace this head with a state-of-the-art servomotor operating under a new digital control scheme to drive a brand-new type of lead screw device to produce linear motion.

The team was not sure yet that this new technology would work, and they had some motor overheating and burnout problems before arriving at an acceptable design. The big question, however, was whether all the new hardware would survive and maintain its alignment for the millions of cycles required in the spec. The engineers had to get a servomotor to run long enough to accumulate the cycles needed for an adequate test of parts wear. Because this was the key technical risk in the machine, they maintained an updated design of their traditional pneumatic head and kept it ready to slip in if they reached an impasse with the servo head. Their product would still be popular even with the old pneumatic head, but some prospective customers would no doubt object to its noise. Then when enough cycles had been accumulated on the new head to know that they had a winner in it, they stopped work on the original, pneumatic head.

The second example comes from the Phoenix refrigeration unit developed by Carrier Transicold for truck-trailer applications. This project was completed in just six months, compared with the eighteen to twenty-four months normally required to develop such units. The major innovation in it was a stylish all-new housing that later won a top award from the Industrial Designers Society of America. Its rounded panels are made of bonded metals, which was at the time a brand-new technology for Transicold and one at which its competitors had failed in attempting to incorporate it into their own products. The Transicold team was eager to make the new technology work, but the new adhesives, which were developed in conjunction with 3M, failed to bond adequately until the very end of the project. All along—for the full duration of the project, in this case—the team maintained a welded-panel fabrication approach for use as a backup if necessary. This approach would have increased production costs and disappointed the technology advocates on the team, but would still have gotten the product out on time.

In Weak Areas, Model and Test

Designing can either be done on paper or its equivalent, the CAD screen, or be done by building models and testing them. The paper approach would at first glance appear to be the more sophisticated, faster method, but paper representations of parts are poor models of reality. They may be helpful in showing dimensional details but hide other problems, and drawings cannot be tested to see if they work. Consequently, there is a great deal to be learned from making quick-and-dirty models and testing them, which often allows the team to make progress faster, because problems are exposed quicker.

To be sure, designing on paper is still needed, but in most organizations the balance needs to swing more in the direction of the empirical, especially when speed is the driver. Making models cannot be allowed to become a substitute for shoddy thinking or analytical laziness. The astute engineer knows how to exploit and combine empirical methods with analytical techniques, to learn at the quickest rate. The point in stressing model making and testing is that in most companies simple models are frowned upon, there are no facilities where engineers can make and test things themselves, and the central model shops and laboratories produce the highest quality results but are typically backed up with work for months. Model making is often a major queue in the development process.

There is simply no substitute for making something and seeing how it works, feels, or looks. As one of our clients who formerly worked for a chain-saw company once put it, "At some point, you have to cut the wood."

David Kelley, one of Silicon Valley's product design wizards, preaches "quick prototyping" to cut rapidly through product design complexities. As a professor Kelley teaches product design at Stanford's engineering school, and as the head of his own design firm he has created the mouse for Apple Computer, as well as over two hundred other innovative products, from telephones to medical diagnostic equipment.

Kelley designed the Apple mouse by building dozens of models, discarding most of them immediately. He fashioned the first mouse model from a plastic butter dish. To get the shape right he made about a hundred nonfunctional forms, to check for things like feel and the number of buttons to use. Kelley maintains two model shops for his design staff, one of them exclusively for the engineers' own use.

The Kelley technique is to move quickly and aim the models precisely. As he puts it, "If we took a year to go through each iteration, no one would think we were very clever. The trick is knowing the critical features to be tested with a proof-of-concept model, while ignoring superfluous details that would only slow down the prototype construction."

In addition to accelerating the development process, Kelley uses quick prototyping to reduce both technical and market risk, by putting real hardware in customers' hands. As he explains:

> Everybody has seen cases of poorly designed products that make it all the way to the end of the development cycle before the company realizes that the product fails to meet basic user requirements. Automobile oil filter wrenches used to be a good example of this—they would break before you finished your first oil change. Our iterative design methodology would uncover that kind of weakness in the first or second round of model making, and the product that reaches the market would be a third- or fourth-generation product that really *works*.
>
> Three-dimensional models are irreplaceable for generating quality client feedback on a product during development. For example, when we discuss product concepts or present drawings, clients typically make vague statements like "This seems about right." On the other hand, that same client might look at a simple model of a computer tilt mechanism and say, "The range of motion side-to-side is fine, but the up-and-down movement seems to be too limited." There is no substitute for that kind of quality feedback, and we feel it far outweighs the cost of a quick prototype. (David Kelley Design, personal communication).

To use models in a risk-control mode, first identify the high-risk areas, then build models or design experiments intended to resolve the risk as quickly as possible. Strive for the very simplest models or experiments that will answer the question. If there are two questions, build two different models, which will likely be less complex than a single model built to test two separate ideas. And always try to test at the lowest possible level when dealing with components or subassemblies.

Simple models are also desirable for another reason. It is always tempting to build a complete prototype of a product, to see if it works. But if it does not, you have invested a lot of time and made one big mistake. Instead, learning accrues faster and risk can be managed better by making lots of little mistakes.

Models and testing are appropriately used to resolve either technical or market risk, but avoid having models that attempt to do both. Models to test technical concepts can in effect be breadboards that are not attractive or user friendly. Market-test models must usually have facades that look like the desired product, but they seldom need to be at all functional. For instance, Maytag once tested a stacked-appliance concept simply by taking wooden mockups to shopping malls and getting consumers' reactions to the stacked configuration. What they learned from plywood enabled them to make wise decisions that greatly increased the popularity of their stacked washer and dryer combination.

SUGGESTED READING

Peters, Tom, and Nancy Austin. 1985. *A Passion for Excellence*. New York: Random House. "Part III: Innovation." Fascinating stories of how inherently messy the innovation process is, not being amenable to conventional management or risk-control measures. Peters and Austin stress that only through constant experimentation (i.e., small wins) can one make progress.

CHAPTER 13

The Product–Process Duo

The relationship between design engineering and manufacturing engineering is undergoing rapid change in manufacturing firms. Just a few years ago people lightly spoke of "engineering" throwing drawings over the wall to "manufacturing," and we were using the phrase "getting the design into manufacturing." This kind of language is no longer adequate.

The use of such techniques as design for assembly precipitated the first phase of this transformation. Having these techniques soon made product designers aware that they should pay close attention to what would likely happen to their drawings on the other side of the wall. But as things turned out, that was not good enough—manufacturing people actually had to get involved in the product-design process.

Organization charts began to change as design engineering and manufacturing engineering started to get closer together. Because these charts can be easily rearranged they do not often say much about the actual workings of a company, but then new course titles such as simultaneous engineering, concurrent engineering, and integrated engineering began to show up on the seminar circuit. Companies struggling with these issues of course realize that terms like "simultaneous engineering" miss the point too, because what is really needed is simultaneity in everything, including purchasing, field service, finance, and quality.

With the terminology in turmoil, we have chosen this chapter's title to emphasize that a product's design and the associated manufacturing process for it must evolve simultaneously. In simple terms this is a major opportunity to save time, by overlapping the product- and process-development activities.

THE OPPORTUNITIES AND THE DIFFICULTIES

There are two ways in which product–process design integration can save development time. One, as just mentioned, is the possibility of overlapping two major tasks that have traditionally been done sequentially. The other is that working on the product and the process simultaneously typically results in superior design trade-offs that avoid the substantial downstream delays that result from poor decisions.

Much of the work done to date in integrating product and process design has been driven not by time-to-market concerns but by considerations of product cost and quality. After trying ineffectually to make changes in an existing production process to cut cost or raise quality, people have discovered that the only way to make fundamental improvements in either objective is to start at the product's conceptual design stage, where the basic design choices are made that will influence the manufacturing processes and therefore a product's cost and quality.

These initial process decisions, which are often made unknowingly, also have a profound impact on development schedules. From a speed perspective it is an issue of doing things right or doing them over. If the proper process decisions are not made at the first opportunity, the product design nevertheless takes form and solidifies around the wrong process alternatives. The later this mistake is discovered, the more costly and difficult it is to correct. Once it has been discovered, we have two options. One is to live with the poor decision, which will have product cost and quality ramifications over the product's life. The other is to go back and correct the initial mistake, which is likely to require substantial redesign and lost time, both being schedule issues.

There seem to be no data that directly show the impact of initial process decisions on the length of the development cycle, but indirect evidence suggests that the impact is substantial. Several organizations have themselves calculated the percentage of a product's manufacturing cost that is attributable to choices made during the product design stage. A recent McKinsey & Company study of an automobile body panel found that of its 100 percent manufacturing cost differential as

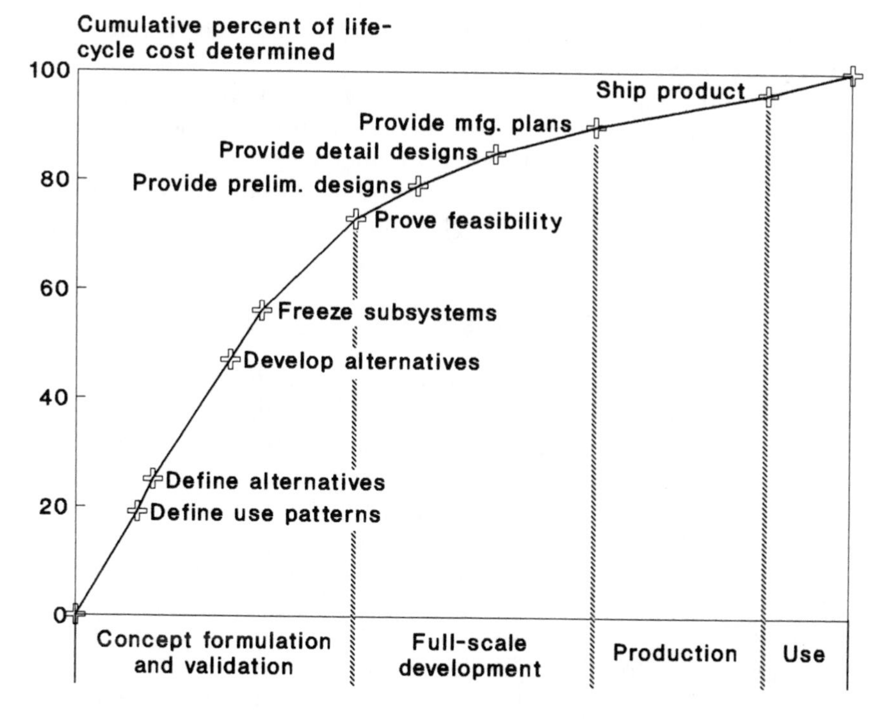

FIGURE 13-1.

Product life-cycle costs are locked in at an early stage of development, when fundamental design choices are made. These choices, which include process issues, also influence a product's development time. *From Nevins et al.,* Concurrent Design of Products and Processes, *McGraw-Hill, Inc., 1989, with permission.*

compared to a competing panel, three-fourths was created by design choices. And a British Aerospace study has reported that 85 percent of a product's manufacturing cost will depend on choices made in the early stages of its design. Rolls-Royce investigated two thousand components and found that 80 percent of their production cost was attributable to design decisions. These findings are summarized in Figure 13-1, which is a composite of data from companies such as Ford, General Motors, and Westinghouse. "Early" manufacturing involvement often means at the stage when preliminary designs are available for review. As Figure 13-1 indicates, 80 percent of a product's life-cycle cost has already been determined at this point.

These high percentages suggest that important process decisions are not getting made correctly at the product design stage. Furthermore, these cost differences reflect only the poor design choices that made it to final design—an additional group probably had to be reversed, resulting in redesign work and delays.

Thus, arranging for early manufacturing involvement provides two types of opportunities: chances to overlap product and process activities so as to shorten the overall development cycle, and ones to minimize redesign work stemming from product design choices with unacceptable manufacturing ramifications.

Process Design as an Orphan

The statistics just discussed are a consequence of the fact that process design simply is not getting done in most companies, at least not at the stage of development where it has maximum influence on a project's outcome. The companies concerned do have process engineers. The problem as we have observed it is that while the product engineers are making crucial design decisions, the process engineers are typically out on the production floor solving ongoing production problems or trying to clean up the mess of introducing the last new product. The reason the last new product is a mess is that the process engineers were busy out on the floor with another product when the crucial design decisions were being made for it. Process design has thus become an orphan, because process engineers' involvement is out of phase with the need for their services.

The time a process engineer has available for new products is often diminished greatly by the demands of ongoing manufacturing troubleshooting for existing products. This is particularly true when a company is upgrading its production equipment or installing new systems, such as JIT. Even without these distractions, however, a process engineer's allegiance is often more toward production than toward new products. Production problems have a natural urgency that product development cannot match, particularly when the process engineer works for the plant manager. In addition, in most of the companies we have visited there is a priority system, whether explicit or implicit, in which today's production always comes first. Other members of the product development team can escape the pull of the daily numbers, but the process engineer is likely to be snared by this trap.

Another type of paradox also faces the process engineer. On the one hand, this person is expected to be involved with the design team at a conceptual level, providing input on design producibility. On the other, the process engineer needs to be out in the factory to learn about the nitty-gritty issues that will make him or her valuable to the design team. We have seen process engineers who have been moved from factory to design lab and after a few years lost their ability to provide process expertise. To be most effective, process engineers need to spend a significant amount of time on the floor, sharpening their process skills, but they also need a degree of immunity from production problem solving.

Companies often regard process engineers as second-class employees, then wonder why process design is an orphan. Two quite different roles are demanded of the process engineer. One, which is similar to that of the product engineer, requires broad conceptual skills for system design, risk identification, the balancing of trade-offs, and long-range planning. The other role is a shorter-range problem solving and implementation mode. Some engineers do not have adequately developed skills in both areas. In the case just mentioned, the person's implementation skills withered, from isolation in a product design environment. The opposite also occurs when, frequently after many years of production firefighting, a process engineer is discovered to be ill equipped to participate at the conceptual level. In short, process development is a demanding job, from both the breadth and the depth it requires; it therefore must be filled with well-trained people.

Paperwork and Control

The "toss it over the wall" system may be slow and fraught with various other problems, but it does have the virtue of being clean. In the pure form of this system drawings enter under formal control just before going over the wall to manufacturing. Before this, manufacturing knows that anything on the drawings is still subject to change, so they regard them suspiciously. They are reluctant to spend time on the drawings or do anything important with them, such as designing associated fixtures or discussing them with suppliers.

However, when manufacturing is involved in the development process early, they will be working concertedly with drawings that are, at least initially, under minimal formal control. Now, drawing control and the release to manufacturing are no longer synonymous. The types of informal and soft controls described in Chapters 9 and 10 prevail. This not only requires new discipline on the part of product design to keep manufacturing supplied with current information but also dictates a whole new operating style for process engineers who, due to past experience, have become suspicious of unreleased drawings.

This approach also opens the whole question of when drawings should come under control. There is a great deal of variation on this point. The best systems have the flexibility to trigger different controls selectively for different parts at various times, depending on the need for stability as against the need to make further changes. A forging drawing for a long-lead-time supplier may come under revision-level control well before its associated machining drawing does, for instance. Ultimately, a project entering pilot production with uncontrolled drawings is likely to have problems. However, in a highly overlapped process drawing control is likely to come relatively late, especially from manufacturing's perspective.

Where to set the drawing-release control point deserves some experimentation. Some companies place their drawings under control too early, then find themselves burdened with lots of slow-moving ECNs (engineering change notices). Others wait too late, and as a result have problems with the plant or with suppliers making parts to the wrong drawings. The release point for a part should be shifted to balance out these two outcomes.

Interestingly, such computerized control systems as CAD and those for bills of material can either liberate the team by facilitating monotonous tasks and enhancing communication, or stifle it by imposing limitations at every turn. For example, it is common for companies having a computerized common bill of materials system that require a part to be "in the system" before it can be ordered, even in prototype or pilot quantities. Waiting for a clerk to assign part numbers can thus hold the entire development process hostage. Such controls do make purchasing's job safer, but they slow down the building of prototypes and pilot units. It is therefore up to management to make sure these systems serve their needs.

Early Manufacturing Involvement

The complications of getting manufacturing involved in product design early are far outweighed by the benefits to be gained from it. Many companies are now shifting to this mode of operation. For example, Dynapert first tried assigning a full-time process engineer to a pilot accelerated-development project that was intended to cut development time in half. It took them a little time to get used to the new mode of operation. At first the process engineer was not sure what he could contribute so early in the process, and the assembly floor kept trying to tug him away from the team. Then the product engineer did not know quite how to interact on process issues so early in his thinking. Before long, however, the two of them were working out the concept for a part on a chalkboard before it got to a CAD terminal. Now this company routinely assigns a full-time process engineer to be an integral part of every development team.

In addition to Dynapert, it would also be instructive to examine General Motors because their products are both complex and are produced at high volume, with abundant product–process complexities.

GM's size also provides them the resources, however, to study their own development process intensively and experiment with some novel solutions. Let's see what can be learned from GM's approaches that might apply as well to smaller development projects.

General Motors's size dictates having a more complex team structure than smaller companies need to design its products. GM's standard practice is, however, to assign both product and process engineers to each of their product-management teams, which are groups of about a dozen people responsible for developing a certain subassembly, such as a car's doors or braking system. In this way GM has product and process experts working side by side at the lowest level of its product-development team structure.

By integrating product and process expertise at low levels and making associated changes in the product development process GM has been able to shorten their development decision-making process substantially. To illustrate the savings, consider the method of designing a taillight assembly, contrasting the old development process with the current one. In the old process, styling took the lead by creating a light to complement the remainder of the car (see Fig. 13-2). Engineering and manufacturing would then review styling's concept and recommend changes for function and manufacturability. After an agreement was reached, engineering would go to work—but in three groups, each working relatively independently. Lighting engineering would design the optics and other functional aspects of the light and its mounting. Body structure engineering would design the surrounding sheet metal. The third group had to design the bumpers to protect the lights from low-speed-impact damage. When the engineering of the product was complete, a review was held of the entire design, and after any revisions needed to accommodate manufacturing's concerns it was released to manufacturing engineering.

In the new process, shown in Figure 13-3, many of the design iterations that originally wasted time in the styling review are now addressed early on by having product and process engineers supply styling with design criteria (as, for example, setting the overslam clearance between the trunk door and the light). Now, after styling completes a clay model, the lighting, body structure, and bumper teams hold a joint meeting, which goes much faster than the earlier styling review because styling has been working according to engineering's criteria. Then the three development groups design the taillight and its manufacturing process simultaneously, with substantial interaction between the groups: regular interface meetings plus problem-solving sessions, as needed. This flow creates a design to which all parties will have concurred at some point along the way, so there will be little need for a final review.

Having early manufacturing involvement provides three basic

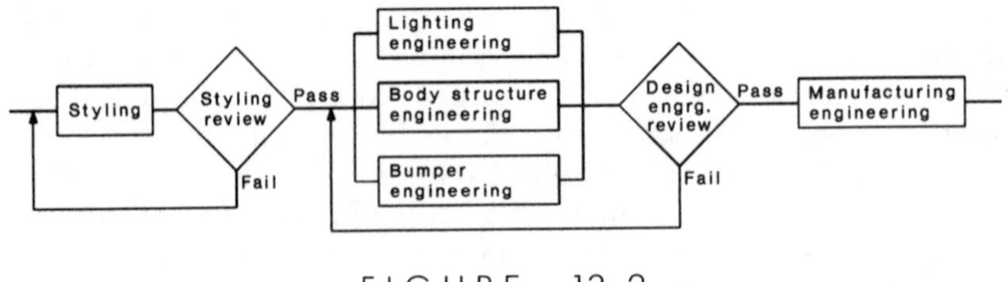

FIGURE 13-2.

The development of an automotive taillight, using an older, sequential process.

benefits: more overlap of tasks that used to be done sequentially, better design trade-offs by aligning designs with production capabilities, and early highlighting of potential problem areas. Process engineers know the company's manufacturing capabilities well and are in a position to review an evolving design, to make the best use of the firm's manufacturing strengths. An experienced process engineer will know the strengths and weaknesses of the plant's equipment and be able to help steer the design in a direction that will minimize downstream problems. At times a product's design will require something beyond existing manufacturing capabilities, but then the process engineer is there to spot the conflict and apply special attention to it, by suggesting a product design modification, extending existing capabilities, or locating outside capabilities to handle the special requirements. In any case, this area of technical risk gets identified early and accurately, so that problems can be resolved without jeopardizing the schedule.

When manufacturing is involved early, it becomes easier to manage the workload as the drawings move into manufacturing. In contrast, with an over-the-wall approach we see what has been described as an elephant traveling down a boa constrictor. What typically happens

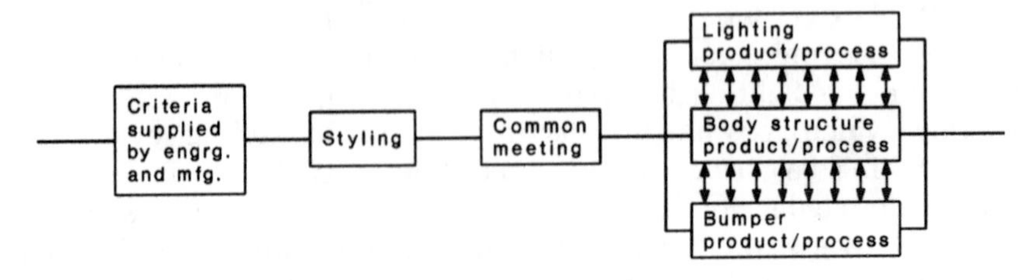

FIGURE 13-3.

The development of an automotive taillight, using a simultaneous product–process design approach.

there is that product engineering holds drawings for most parts of the product until they feel sure that everything is in good order and that all the parts will fit together. Then a whole pile of drawings is released into the stream of manufacturing's paperwork. We recall seeing one project with over six hundred new and unique parts, and over three-fourths of these drawings were signed by the chief engineer and released into manufacturing in a single day! That project was easily tracked just by watching the pile of drawings move from desk to desk as manufacturing attempted to process them. By involving manufacturing in the design process, this glut of paperwork can be planned so that it moves through the process more smoothly.

FOSTERING INTERPLAY

The essence of early manufacturing involvement is that it enables a process engineer to interact closely with a product engineer during the product design phase, especially during its formative conceptual portion. This is easier said than done, unfortunately. What typically happens is that the process engineer is busy at the start of a project cleaning up the details on the last project, as noted earlier. It does not seem critical to have the process person there during the first few weeks because there is apparently nothing to be reviewed. So the organization compromises and allows the process engineer just to attend team meetings while the team is getting started.

This simply is not adequate. For one thing, design details usually are not discussed in team meetings, so the process engineer will not be exposed to the design concepts that are beginning to evolve. Moreover, his or her creative energies are no doubt still being consumed by problem solving on the last project. Rather than doing some what-if thinking to project the outcomes for the new project or perhaps experimenting with a variant of an existing process to suit the new product concept, the process engineer will more likely just be tucking the new project's team meeting into an already busy day of troubleshooting. This person's allegiance is still clearly to the old project. In a few weeks or months it will shift to the new one, but by then the key product and process decisions will already have been made. It will now be more a matter of execution than of innovation for the process engineer, and the same old cycle will repeat itself.

The first time through this new loop, the process engineer will need extra time at the beginning because this is essentially a new task assigned to the organization. There will be no precedent for it, and no job description. It will not be an easy job because its first task is to win the confidence of the product engineer.

The first decisions made about a product form a foundation upon which more and more detailed decisions will eventually build a complex, interacting maze. For example, the first decision on a particular part may be whether to make it of plastic, aluminum, or steel. This is a product design decision because these materials have different strengths and other design attributes, but it is also a process decision. Each material requires a different primary manufacturing process, each of which leads to different streams of secondary processes, such as drilling and tapping. On the product side, one of the later decisions may be about corrosion protection, which will depend on the material chosen in the beginning. If steel is to be used, its processing is likely to require sending the partially completed part out of the factory for plating, which may destroy an otherwise beautiful opportunity to use JIT production for the part. In this simple case one is faced with three unpleasant options. One is to go ahead and plate the part and incur penalties in production cost, inventory requirements, and lead time. Another is to shift from using carbon steel to stainless steel, which will require only minor part design changes but dramatically increase the raw material cost. The last option is to go all the way back to the beginning and redesign the part from aluminum or plastic, which will disrupt the schedule substantially.

The cost of correcting an initially poor decision gets higher and higher as a team gets deeper into its design. The problem will be often resolved with a brute-force solution that increases manufacturing cost. It is these decisions that give rise to the statistics cited at the beginning of this chapter for the percentages of manufacturing cost that are determined by product design. Sometimes through, a manufacturing cost increase will be deemed unacceptable, and often the problem involves product performance or quality issues. In such cases the team often has no choice but to go back, redesign the product, and pay the penalty in development time. The way to avoid this trap is to apply the best process expertise available to the crucial initial decisions.

In addition to making use of the opportunity to draw process engineers in at the beginning of the design, there are other types of manufacturing expertise that should be involved. For example, General Motors first decides where a new car is likely to be built, then invites five seasoned hourly assembly workers from that plant to become members of the design team, to ensure that the design is buildable. For example, Mike Milligan left an assembly-line job at GM's Wentzville, Missouri, plant to become an assembly advisor on the 1991 Buick Park Avenue development program. He temporarily moved to Flint, Michigan, for twenty-four months, where he drew from his twenty-one years of auto-assembly experience to answer dozens of questions a day on car-assembly methods.

Here is one example of the way in which this man's expertise helped GM correct an assembly problem while it was still relatively easy to fix. The engineers had done a crash test on a prototype and found that its front-end structure crumpled a little differently from what they had expected. The steering column was thus pushed up higher than planned, so that the air bag in the steering wheel hub overshot the simulated driver. The engineers' planned correction was to be a brace connecting the towers above the two front wheels. This seemed a simple way to reinforce the front end, but Milligan immediately saw a problem: although the brace was designed to fit around the engine after the car was assembled, it would obstruct the path for loading the engine into the car. With this information the engineers redesigned the brace to allow smooth assembly. Milligan's years of assembly experience provided knowledge that otherwise was missing on the design team, even when it included process engineers.

There are two approaches to product–process integration that are usually not effective if the objective is to accelerate the development process. One is the ultimate in integration: to combine product and process expertise into one person. These are two broad and demanding roles, however, and combining them may well result in getting neither the best of product design nor good process design.

The other approach is to provide an integrator or liaison to connect the product and process activities. This will prove to be an inadequate substitute, however. These two functions must work with each other intimately, and placing an intermediary between them will only compound the communication problem.

Getting Product Engineers onto the Floor

So far we have emphasized bringing the process expertise to the product designers. The product designers also need to get some hands-on experience with manufacturing processes, though, so that they can develop an appreciation for how their design will be manufactured.

As consultants we watch for bottlenecks in the development cycle and sometimes find that the process slows down when it gets to the manufacturing floor. We then tour the factory and talk with some of the people associated with the slow products. Amazingly, we often find that these people have no idea who the product engineer is! A design engineer's project may be unraveling on the factory floor, but the engineer has yet to appear on the floor to deal with it.

Several benefits result from instituting product engineer contact with the fabrication and assembly people. One is that the engineer can now get a feel for what it is that makes products hard to manufacture. It may now be too late to apply much of this learning to the current product, but it will help the next one slip through manufacturing with fewer hitches.

The other benefit is more a psychological one, but is also powerful. Manufacturing tends to work on a short time scale, compared with that of engineering. Their job is basically to get today's orders out today. Consequently, they can be expected to have limited patience with a new product—which no one has even ordered yet—that cannot be manufactured according to instructions. If the factory is busy, they will just set the new job aside and get on with the "real" jobs. They will surely work on the products they are measured on—today's products. However, every time a new product is set aside the project is delayed, often for quite a considerable period, because it can take weeks to get the new product scheduled back into the production stream.

A new product really needs a champion in the factory. Product engineers are ideal for this role because of their broad interest in and knowledge of the product and as a result of their authority to make changes in the product if necessary. By contrast, process engineers are too close to manufacturing's problem-solving cycle at this time to let them be effective champions. The product engineer can just wander through the plant, explaining to key workers why the new product is important to their futures, how the engineers have tried to make it manufacturable, and acknowledging that there are still some open manufacturing questions for which their help will be needed in order to resolve in volume production. With this kind of preparation, manufacturing's workers are more likely to be in a tolerant, problem-solving mood when working on the new product, rather than feeling resentful or impatient.

General Motors goes further than just having its engineers wandering around in the factory—it puts them to work building cars. Design engineers from GM's Flint Automotive Division spend a full day each quarter assembling the part of the car that they normally design. They thus learn some valuable lessons that will eventually help streamline the initial assembly of new car models, lessons that cannot be learned from books. For example, when trying to push some parts like hoses onto a car as it is continuously rolling down an assembly line, it is much easier to push the part on from the side, because if you have to work from the rear, the target constantly recedes as you are trying to bring the two parts together. Designers also learn that threaded fasteners can be difficult to install quickly, particularly in tight places where it is impossible to get all of one's fingers on the nut to get it started. As these engineers learn more on the assembly floor, we will probably be seeing fewer nuts and bolts in GM cars.

Some of this knowledge can be and has been codified and standardized. Many companies already have lists of standard parts (electronic components or fasteners, for example) that they encourage their design engineers to use. The use of standard parts may not provide the most elegant or inexpensive design, but it does minimize inventory requirements. More importantly from our viewpoint, it reduces technical risk to use a proven, familiar part in place of something that is either newly designed or even new to the company, and it generally helps support quicker schedules.

Another way of codifying manufacturing processes for product engineers is to conduct what are called process capability studies on the different processes in the plant. These analyses provide guidelines to design engineers on what can be done efficiently, keeping up high quality levels, in the plant. For electronic assembly, for example, depending on the type of machinery available one can specify the spacing, orientation, and shape of parts that can be assembled automatically on printed circuitboards, as well as giving the other parameters of boards that will fit the machinery. For mechanical fabrication such a list might include the tolerances that can be held, the part sizes, and so on. These guidelines are to help designers avoid unexplored territory in the factory, which then speeds up manufacturing's debugging process. Instead of attempting to restrict design engineers, these guidelines remove mundane issues and help them focus their creative energies where they will pay off in the marketplace.

Building a New Product

One of the last steps in bringing a new product to market is learning to assemble it at volume rates under actual factory conditions. Often this step is assumed to be trivial, until the development team is presented with surprises at the tail end of the development cycle, which then leads to unexpected delays in introducing the product. We have seen this occur in many different industries, but some companies have found ways of overcoming the problem.

GM, for instance, has huge challenges in getting an assembly line to work smoothly, so they have created a bold solution. Formerly, they built a pilot lot to test the assembly process, but it came too late in the development cycle to make corrections before regular production began. Also prior to the pilot run, they made some prototypes in the laboratory for engineering testing, but these lab-made prototypes provided no information about potential production assembly problems. So GM added another construction step: production-line prototypes.

These cars are made from prototype rather than fully tooled parts, but they are built early, using an actual assembly line. Through careful planning GM does not have to shut down an operating assembly line for more than a few hours, and they can arrange to make a car in steps, so that a whole assembly plant is not shut down at once. This early experimental build provides a lot of feedback that helps modify both the product and the process design. These cars then remain at the assembly center for later use in training assembly workers to do their new jobs. Because these prototypes cannot be sold to the public, this is yet another example of what we have called building a tall junk pile, by trading money for time.

By contrast, Dynapert does not produce at high volume, so assembly problems are less critical to them, but they have had some difficulty servicing the initial units of a new model in the field. Their solution to this start-up problem has been to call in a field-service technician to be part of the development team and help assemble prototypes, rather than using the customary laboratory technician. This step enables the field technician to spot possible serviceability problems while there is time to correct them. It also provides a field technician who understands the machine once it is launched, who can service the machine and also train other field-service people. Simultaneously, the company thus solves some design-for-serviceability problems and trains its service technicians.

We should not leave the impression that only big companies have big assembly problems. One company that we know of makes sewn fabric goods for the consumer market. They have a custom seamstress who can work wonders with fabric. The catch is, of course, that their piecework operators cannot replicate her methods reliably under volume production conditions. After struggling with the original designs provided, the assembly foreman has to redesign each item so that it can be manufactured, a step that was not allowed for in the initial schedule or budget. Even in small operations making simple products, manufacturability needs to be designed in from the start, or it is likely to delay a project just when management believes the product to be ready for production.

PLANNING FOR PRODUCTION

Actually getting a product made can be considered more of a tactical issue. The related strategic issue is that of providing production resources for the new product in a timely way. Some of the material in

Chapter 11 on allocating resources applies here, but there are also additional points that pertain to planning production resources.

This area is one in which top management needs to be involved in understanding what the team needs and be ready to support it. Many of these production-planning issues fall outside the authority of the team.

The point we have been stressing throughout this chapter is that manufacturing processes need the same kind of attention to development that the product itself customarily gets. However, the scale of the two activities makes it difficult to fund manufacturing development at the same depth as product development. A product can often be built and tested using just small-scale or partial models and jury-rigged test setups. On the other hand, testing the process can mean buying expensive machinery, building full-scale facilities, and consuming lots of parts. Therefore, manufacturing development requires particularly thoughtful planning to exploit the available opportunities and use capital wisely.

Consider how one company that makes machined metal assemblies integrated their manufacturing development process into their product development. This firm wanted to get into the high-precision end of their market, and were fortunate that designing a new line of high-precision products was relatively straightforward. But then they had to learn how to fabricate the new precision parts, a task beyond the capabilities of their machinery. The conventional procedure would have been to design the product while preparing a request for capital to procure the high-precision machinery, order the machinery, wait for it to arrive, install and learn how to use it, and then—after the design had been sitting idle for months—refine the manufacturing process until it could make production parts.

However, this company took a different view of this opportunity, with a different sequence and outcome. A recent acquisition had already established that they would be getting into the high-precision end of the business. So rather than justifying the new machinery based solely on needing it for the new product, they bought it because it was part of their new business, even though they had no compelling need for it right away. Then when they did get it set up and running, they immediately started making parts of the best precision possible and put them into their normal products. They were in fact giving away precision because these parts cost more to make than those normally used, but they were learning, preparing for the new product's arrival. When the product design was finally ready to manufacture they had a stable, high-precision manufacturing process ready for it, with no start-up time needed for the new machinery and very few scrapped parts.

Machined castings are an example of parts that have complex manufacturing processes, usually with a long list of operations, and special tooling and fixturing requirements. These manufacturing processes usually do not work satisfactorily the first time—they have to be tried, adjusted, and retried in an experimental mode. This takes time, which is often not in the schedule because planners tend to optimistically assume that a new manufacturing process will be well understood and work properly the first time. It is more realistic to assume that the process will not be acceptable initially and plan to have to change it. If this extra time is planned into the schedule it need not stretch a project, because it can often be done concurrently with other activities.

Getting early manufacturing experience with a process, however, can mean doing things in an unorthodox way, which can create problems, unless the rationale for them is clear to all involved. For instance, Chapter 9 cited a case in which manufacturing got its early experience by machining defective castings, deliberately creating scrap. Though this may be a good business decision, it will prove a difficult concept to sell to a plant manager who is measured on his or her scrap rate.

Thus, the logic of getting the manufacturing learning process out of the way early may be clear to the chief engineer, but it also has to be made clear to the machinist. Machining scrap will send confusing signals to machinists who have probably been steeped in zero-defects teaching and don't understand the game plan.

There is another area in which management can inadvertently send mixed messages easily. This area involves the changeover from an old product to its new replacement. In order to protect themselves from development schedule delays, management often produces a substantial safety stock of the old parts. Then when the new product is ready for production, the production-planning system will not need it

yet, because it needs to work off the supply of existing parts. It is then difficult to build a compelling message for marketplace urgency when the company is actually operating in a hurry-up-and-wait mode. Here again, JIT teaching can be transferred to new product development. Rather than maintaining its safety stock, the organization needs to start lowering it gradually and simultaneously solve the development cycle problem so that new product delivery becomes more reliable.

The high-precision illustration just given is an indication of another manufacturing-planning pitfall. Corporate financial planning systems often drive a project into an increased level of complexity (see Chapter 4 for a description of the detrimental effects of complexity) by encouraging operations to change the manufacturing system along with a new product. Having a new product can be an opportunity to upgrade machinery that needs replacing anyway, and the corporate capital-approval process can make this the easiest way to effect the factory changeover.

From the point of view of minimizing complexity and thus speeding up product development, it is usually best to plan on not changing over the factory with the introduction of the new product. Instead, schedule factory changes to occur either before or after the new product so that when the inevitable manufacturing start-up problems do occur they can be traced to either the new equipment or the new product, rather than an unknown amalgam of the two.

Finally, remember the users of the new machinery while still in the planning process. Without grass-roots enthusiasm for new machinery, it can take much longer to render it operational. Duracell, the worldwide battery manufacturer, provides a good example of how to build this enthusiasm. The batteries made by Duracell employ deep-drawn metal cans to house the chemical ingredients. As a major producer, they are often approached by machinery manufacturers offering new, better deep-drawing machinery. Duracell's director of product engineering, Terry Eisensmith, was once impressed with the apparent advantages of one of these offerings, made by Platarg Engineering of Great Britain, but knew that the people on the plant floor had to be impressed too. They tended to be skeptical because they had seen many promising new machines that had created production problems. Eisensmith therefore included the production foremen in the purchase decision.

On the first trip from Connecticut to England to assess Platarg's machine, Eisensmith took along the first shift foreman. Together they were impressed with the machine's capabilities, but before its delivery a second foreman went to England for a demonstration, and a toolmaker responsible for machinery setup went to observe the acceptance trials. When the machines finally arrived the start-up period went quickly, because the key implementers had already been sold on the machine's potential, and with their personal relationships with Platarg personnel they could iron out any remaining problems by calling their contacts in England. This is yet another example of trading money for time.

THE BROADER CONTEXT

The ideas presented in this chapter can have a profound impact on time to market, because of their ability to remove surprises from the tail end of the development process. Recovering from late surprises, when very little time is left, is particularly difficult and costly.

Remember that the issues presented throughout this book apply to other manufacturers as well, specifically to suppliers, and particularly to suppliers of tooling. If procuring tooling and fixtures is slowing things down, it may be possible to work out some creative solutions with the supplier to benefit both of you, just as many large companies have helped their suppliers institute quality-management programs. The Japanese have found that many JIT methods apply effectively to tool making, as for example the concept of letting machines, not parts, stand idle.

Some of the greatest obstacles to product–process integration can be at the top of an organization. Often the product and process engineers recognize the value of collaboration and would like to eliminate practices that waste their time, such as an overreliance on ECNs, but find the organizational structure and values needed to support collaboration missing. The next chapter focuses on what management can do to accelerate product development.

SUGGESTED READING

Dean, Jr., James W. and Gerald I. Susman. 1989. Organizing for manufacturable design. *Harvard Business Review* 67(1): 28–36. Several thought-provoking options for ensuring manufacturable designs, but these ideas must be considered in view of their impact on development speed. For instance, one suggestion, giving manufacturing a veto power, may eventually yield a producible design but may also waste design time.

Nevins, James L. and Daniel E. Whitney, eds., and Thomas L. De Fazio, *et al.* 1989. *Concurrent Design of Products and Processes.* New York: McGraw-Hill. A wealth of engineering information on product–process design, but weak on fabrication concerns relative to its stress on assembly issues.

Walleigh, Richard. 1989. Product design for low-cost manufacturing. *The Journal of Business Strategy* 10(4): 37–41. Additional discussion of many of the concepts covered here.

CHAPTER 14

The Role of Top Management

Top management's leadership and support are crucial to sustaining any significant improvement in development time. Unless top management is truly interested in faster product development—and shows it—little can be done by lower-level managers and workers to speed up product development.

This chapter explains what top management must do in order to make time to market a company priority and discusses the ways it should, and should not, be involved in specific accelerated-development projects. The next chapter then explains in detail how to make the transition to this faster mode of operation.

TOP-MANAGEMENT LEADERSHIP

Fast product development requires a great deal of concentrated attention. This focus does not exist naturally in a company, though, because individuals or departments have differing goals that are often at cross-purposes. Effective project management and good cross-departmental coordination can do a great deal to smooth over these differences, but until top management communicates the value of speed, the pace of the development effort will be limited.

We often attend conferences and seminars to speak on fast product development. After our presentation, first-level and middle managers sometimes approach us to express their frustrations. They are devoted to their assigned project, but it keeps getting turned on and off, they can't get the resources they need, or the product's requirements keep changing. What can we do to help? Unfortunately, not much—until top management decides that getting new products to the marketplace quickly is worth serious attention, workers will continue getting mixed messages and the firm's development process will be unlikely to accelerate with any consistency. The workers may be ready, but management must decide on the priorities.

Occasionally, a band of dedicated souls will form a skunk works, go outside the regular system, and get something done quickly in spite of prevailing general indecision. This tactic will get one product to market quickly, but it cannot build an ongoing capability. What's more, it probably creates further problems, such as inconsistent documentation, in its wake. In order to build a sustainable, ongoing capability for fast product development, an organization needs to make some fundamental changes in the way it operates, which needs to start at the top.

In working with organizations we typically ask many people in a company at various levels and in different functions why they are interested in developing new products and why the factor of time to market is important to the organization. In doing so we are simply trying to understand the agenda at the company so that we can address the proper issues; what we hear is often surprising. People often are not sure why they should be racing new products to market, or they will have diverging explanations for it. Although the top individual in an organization usually has quite clear views on time to market, these views will not be spread through the organization, so each person will be operating from a different script. Until top management establishes—and communicates—the firm's position on rapid product innovation, the results will remain sporadic and fragmented. (Chapter 3 provides additional coverage of this topic.)

This lack of focus originates because of a discrepancy starting at the CEO level between what is important and what gets done, a discrepancy that is widespread, at least in American industry. Hise and McDaniel surveyed the CEOs of the 1,000 largest industrial companies in the United States. Of these, 236 responded and provided some revealing information on how they deal with product development. They listed product development as the most important of eight sources of future growth and profit for their companies, with allied research and development activities ranking third, together receiving over twice the vote of second-ranked financial planning (see Fig. 14-1). But when asked which functions receive their attention financial planning dominates, receiving attention over twice as often as product development and R&D combined (see Fig. 14-2). In short, management's words support product development—even speedy product development—but their actions communicate a different set of priorities.

Unfortunately, many senior executives have not had the experience of being on the inside of a product development project, so it may be difficult for them to appreciate the dynamics and frustrations involved in bringing a product idea to life. There is really no substitute for hands-on experience on a product development team, but throughout this book we have attempted to provide the flavor of the development process for those who are not experienced with it personally.

Management's Signals

Management communicates the importance of time to market through its actions. Often, however, management's actions or signals are inconsistent, because the management team has not formed a mutual understanding of the value of rapid development. Thus, the starting point for accelerating product development as a corporate pursuit is for upper management to formulate a corporate statement on the role of product development, and rapid product development in particular, in the firm's business strategy. Such a statement becomes especially important if business conditions have recently raised the urgency of getting products to market.

Once the top staff is clear on what they want to achieve with rapid product development, this message needs to be spread all the way to the shop floor and the most obscure corners of the R&D lab. Manufacturing people are usually under pressure to achieve goals that will, to some extent, conflict with fast product development. For example, the factory wall often displays an evolving chart showing its on-time delivery or defect-reduction performance relative to a specific target level. To the extent that production may be rescheduled to allow a pilot run of a new product, for instance, on-time delivery could be affected, and working the inevitable glitches out of a new product may temporarily drive up the defect levels. These constantly reinforced factory messages will apparently contradict the demands of a new product, so management must be particularly clear with production people about the imperatives of getting new products to market quickly. Appendix B expands on this issue.

In some companies the R&D lab is a relatively quiet place where technical people are left to create new ideas without interference from

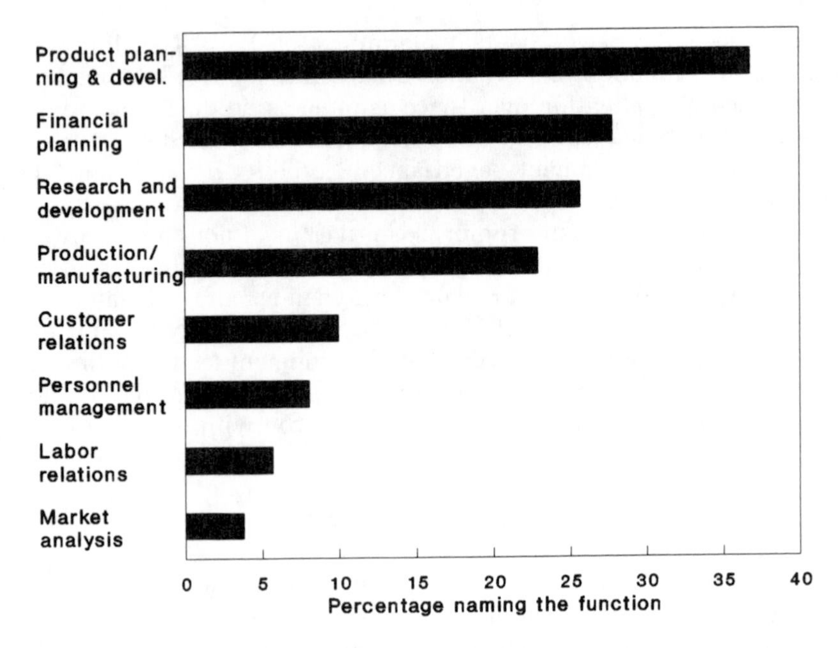

FIGURE 14-1.

The most important sources of future growth and profit, according to CEOs of the largest companies in the United States. *From* American Competitiveness and the CEO—Who's Minding the Shop, *Sloan Management Review, Figure 2, p. 52, Winter 1988.*

customers. Therefore, it can be sheltered from the urgency of the business climate, and technical people may be working on their own, much different, time scale than the one senior management has in mind.

Particularly when an organization is trying to shorten its development cycles, it is essential to provide clear signals about the importance of rapid product development. The schedule itself must take on a new importance; in a sense, it must be worshipped. This is one reason for posting a giant project schedule in the team area, much as charts on manufacturing or safety performance are posted in the shop for all to monitor.

Top management's presence in the various parts of the organization sends amazingly strong signals to the troops about what is important. For instance, some senior executives spend a lot of time with the accountants. Others show up frequently in the research labs or industrial-design studios. As consultants we sometimes ask the development team's members how much value the organization places on having new products. When they give us their perception we ask how they

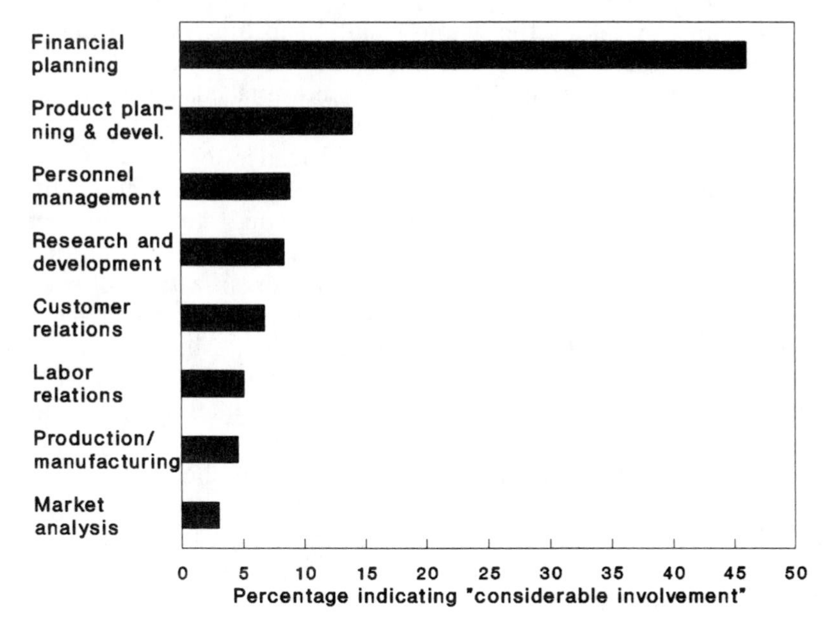

FIGURE 14-2.

Functions with which they have "considerable involvement," according to CEOs of the largest companies in the United States. *From* American Competitiveness and the CEO—Who's Minding the Shop, *Sloan Management Review, Figure 3, p. 53, Winter 1988.*

know, an answer that often has to do with where top management have been observed.

The team's morale is clearly important in high-performance situations like accelerated product development, and upper management's interest, as expressed through its appearances, can make a great deal of difference when team members start asking themselves, "Why are we killing ourselves for this project?" Even executives' happenstance appearances can carry considerable weight. A team leader once said he knew that his project was important to the company because, as he put it, "I ran into the president in the rest room and he asked me how the project was going." This interested executive even monitored progress in the rest room, not by passively scanning progress reports in the office.

Organizations foster what they reward. A presidential commendation for an accelerated project brought in on time sends a much different message than just an across-the-board "bonus." One general manager let the development team know how much he appreciated their contributions by taking the whole team out to lunch when they had reached a particularly important milestone. This created a lasting impression on the team's members, who were unaccustomed to eating with the top brass. (In situations such as these it is also entirely appropriate—in fact encouraged—for the team leader to let management know when something special has been accomplished and what it means to the company. Time-to-market advances are extraordinary; they must be brought to light and be exploited through recognition.)

Because the objective is to transform the organization into adopting a new mode of handling development projects, top management should watch for and encourage desirable changes in behavior. This may not be as easy for management as it might seem, though, as the following example shows. One company that wanted to accelerate its development process assigned a cross-functional team to a project and relocated them into a separate area, as suggested in Chapter 8. The team then took the initiative and decided that they first needed a product specification, so they wrote one, because they had all the engineering, marketing, manufacturing, and purchasing expertise they needed right there on the team. Again, this is very much in line with what we have suggested, in Chapter 5. When the spec was complete, which did not take long, they were ready to dive right in to the design work. Thus, they issued the spec under a cover memo stating that they were starting to design to it immediately and would consider it final unless they received objections within a week.

At this point the political friction entered. The company had always regarded specifications as documents that evolved over months as they floated among departments and executives gathering refinements. In particular, the president, an engineer, and the general manager, who had come from marketing, were normally contributors to product specifications. The executives' first reaction to this quick specification was "Wait; you can't do this so fast; we have to think about it." Fortunately, they quickly reversed this position as they realized that they were getting exactly what they wanted: vast reductions in cycle time. The result was that the executives called a meeting within the week to consider and approve the spec, which was then approved, with only minor changes. It perhaps lacked some of the nuances (and complexity—see Chapter 4) that would have been added in a more lengthy process, but it was good enough. More importantly, the team had already signed up to execute it.

Management was wise here to learn from and support the team in this situation. By nurturing such changes that are at least headed in the proper direction, even if they are not complete successes, management encourages the team to initiate further change. The team and those who interact with them will thus be learning by trying new approaches, which will sometimes result in mistakes. Although managers would rather not have the mistakes, they cannot afford to squelch the learning that occurs during this critical period of transition. Furthermore, the way in which management responds to such changes will send significant messages to the development team regarding the permanence of its time-to-market emphasis. Top management's signals will either encourage further progress or allow the emphasis to wither.

Mistakes unavoidably play an important role in rapid product development. As explained in Chapter 4, product development is a continual process of learning, and to learn requires making some mistakes. To develop products quickly, the goal should be fast mistakes rather than no mistakes. Top management must support and encourage this viewpoint.

SUPPORTING AN ACCELERATED PROJECT

In addition to the leadership that top management provides in establishing justification for rapid product development, communicating the importance of it throughout the organization, and maintaining an awareness of it, there are concrete ways in which upper management must be involved in each particular rapid development project. Let us examine how top management should interact with the team to keep it moving along quickly, enhance its prospects for success, and keep itself off the critical path.

Initiating a Project

Because top management's leverage to influence a project is greatest at its beginning, concentrate management time put into a project at its beginning to get the team off to a good start (see Chapter 3 for further details on this topic). Senior management's leverage basically stems from the product design decisions that have not been made yet. Once these decisions are made, the team will have passed a fork in the road. After this point the team will proceed to make an increasingly interwoven sequence of decisions, all dependent on the one made at that first fork. The farther the team proceeds from the fork, the more costly in both time and money will any changes become. If a senior manager is going to influence the product—say, on its complexity, which determines its development time (see Chapter 4)—he or she will need to set the tone in these initial decisions.

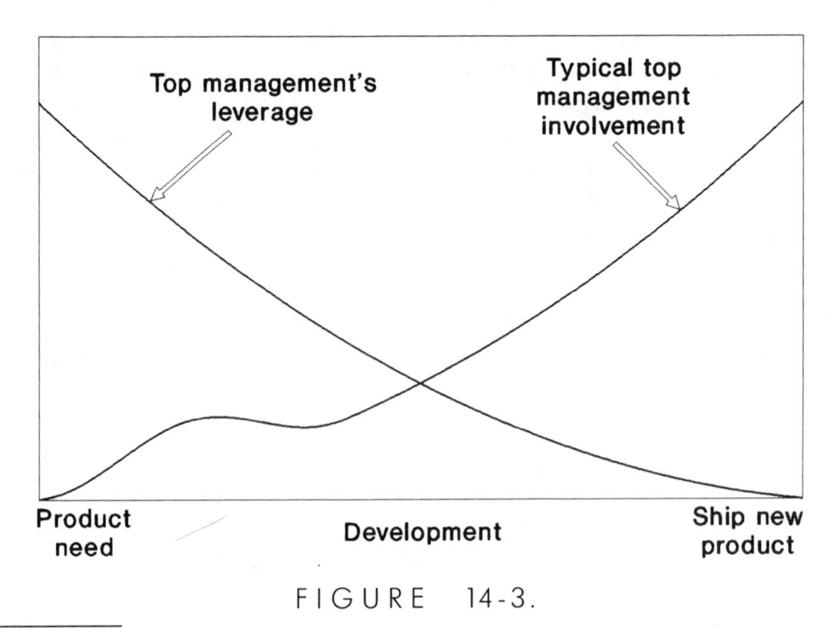

FIGURE 14-3.

Typically, top management's involvement in a project is just the opposite of its leverage in influencing the project's outcome.

Unfortunately, the typical pattern of top management involvement is just the opposite, as illustrated in Figure 14-3. There is normally a small flurry of management activity at the kickoff point, but then upper management often gets heavily involved only after the design is essentially complete and decisions are needed on large financial commitments. As the project moves into production it has a high profile, so it is natural for senior managers to be drawn into any problems that arise at this stage. Except for the kickoff blip, management's typical involvement parallels the project's spending curve, which is basically the inverse of management's leverage curve.

Having top management involved in the front end will also help establish a sense of urgency regarding the project. Most immediately, management must communicate a message of concentrated effort during the project's start-up activities, to overcome the fuzzy front end syndrome that can burn up half the available development time before design activities even start (see Chapter 3). Product and project planning are important, but they can be done much more intensively than they usually are. An appropriate statement from top management at the very beginning can eliminate enormous amounts of delay at the planning stage.

Top management can also help, early in the development cycle, to create clear scheduling urgency by identifying a fixed end point for the project. To establish timeliness as one of the project's requirements, one as important as product cost, quality, or performance, management should use an end point over which the company has no control. This could be, say, Christmas, a trade show, or an OEM (original equipment manufacturer) customer's new model introduction.

Senior management may need to have a part also in selecting the team's leader. As explained in Chapter 7, no other single decision has a greater effect on successfully completing an accelerated development project. Thus, upper management can be influential at the outset of a project by ensuring that a person with a broad enough skill base is chosen for this key role. By selecting a leader with a sufficiently broad perspective top management also sends a message that the leader is to view the project as a business undertaking and make decisions from a business outlook. If management fails to establish this sort of viewpoint within the team the leader is likely to take a technical, marketing, or other perspective. Then key business decisions will have to be made outside the team, which will delay decision making. The economic analysis technique described in Chapter 2 will provide the team leader with helpful decision-making tools.

Establishing an Effective Team Environment

Upper management must provide a team setting in which the team takes full responsibility for the project. There are two parts to this issue. One is to weaken the linkages between the team and the remainder of the organization so that the team can in fact move independently. The other is to create a motivational structure where the team must indeed complete the project on schedule.

Interestingly, these two parts work together, because the best way of helping the team take responsibility for the schedule is to place it in an environment where the external reasons for schedule slippage have been removed. By providing the required resources while establishing a separate identity for the team, upper management encourages it to apply its own resources rather than rely on others who do not have a clear stake in the project's outcome. This type of organizational and physical structure is not common in most companies, so management must have a strong voice in setting it up and maintaining it.

Often the people on development teams perceive that management is unwilling to back up its statements about time to market with the resources to get the job done. Occasionally, team members will test the system by asking for resources, to see if management's statements are hollow. The perception can be quite different from the reality in such a situation, so it is important for management to make its values clear by setting the team up with everything it needs (except extra people on the team) so that the team can look only at itself as deadlines draw nearer and nearer. This may sound like a bottomless pit of demands from the team for more resources, but in reality it is not, once the message gets through.

A few examples will show how this works. One client set up a fast development team and provided it with a small machine shop, a couple of CAD workstations, and a photocopier (this was before the days of the fax machine). There was some initial negotiation over telephones, with the team saying they needed seven and management suggesting that four would do. The team got seven. When the company's general manager made his periodic visits to this team he made a point of asking two particular questions: "How is it coming?" and "Is there anything you need?" The message was clear, and this team came to operate at twice the speed of the organization in general.

The other example comes from the Phoenix semitrailer refrigeration unit developed by Carrier Transicold. Prior to the Phoenix project the corporate values at Transicold had emphasized economizing, at the expense of development time. When management decided they wanted to change this balance, they chose the Phoenix project as the vehicle of opportunity. The company's president told the team leader, Mark Cywilko, "Don't let money become an issue." Cywilko accepted this mission as he watched both for roadblocks and for opportunities to apply resources. For example, he arranged to use the company aircraft once a week for team trips to their West Coast plant and suppliers, clearly a powerful symbolic message. Though this approach may seem simplistic, it worked: whenever the leader sensed that the can-do attitude was slipping, he would say, "Tell me why; I'll get the money." Then he would go to top management to arrange for whatever the team seemed to need.

Besides its usefulness in eliminating excuses, there is another value to this technique. It is really an agreement, a contract, between management and the team. Management agrees to make the resources available to get the job done on time, and the team agrees to meet the schedule if adequate resources are in fact made available. This technique is a powerful one that management can apply to other aspects of rapid product development. For example, the team can agree to meet the schedule if marketing freezes the specification, and marketing agrees not to touch the spec if the team delivers on schedule. This negotiating reinforces the incremental innovation strategy proposed in Chapter 4 in which product planning does not have to be as precise as usual if the product can be delivered before the market shifts. Although the "contracts" to which management is a party will limit its freedom of action, honoring these contracts is a marvelous means of demonstrating its resolve.

The type of separate project structure recommended in Chapter 8 is another issue that affects top management. To be effective the leader of a multifunctional team must report to general management, which can bring disorder to organization charts. Whenever we have worked with management to set up a fast development team the issue has invariably arisen regarding to whom the team leader should report, which can be a substantial stumbling block. This is one more reason management must be careful in selecting a team leader who can operate with a business perspective rather than choosing an engineer who must report to engineering management or a product manager under the influence of his or her marketing management.

Even when the team leader reports to general management, dysfunctional issues can arise from leftover habits instilled by the familiar functional organizational. For example, top management usually understands that the team must be making its own decisions, but in some companies it is hard for functional line managers to let the team run its own project. It has always been the job of line managers to make functional decisions on new products. It is up to top management now to ensure that middle management respects the team's authority. Top management may also have to make this policy clear directly to the team, so that it is in a stronger position to resist encroachments by well-meaning functional managers.

Accounting is another function that often must make some rather radical changes away from its familiar practices, in order to support accelerated product development. We have been pointing out all along, particularly in chapters 2 and 11, that time-based strategies and practices are often different from those that would be chosen by someone with a cost-oriented mindset. More than once, as we have been developing creative ways of "buying time" with product development managers, they have reminded us, "You don't understand. Accounting runs this company!" If the company is indeed going to change from a cost-oriented mode of developing new products to a time-based one, accounting people will need to view their involvement in development projects differently, more as facilitators than as keepers of the purse strings. Because this is likely to be a substantial change in the way they have traditionally viewed their jobs, top management must take the lead in initiating the change in viewpoint and in reinforcing the new behavior.

Lastly, after ensuring that all the changes just discussed are under way, senior management must make some changes in its own involvement in a rapid development project. The basic issue here is to stay off the critical path of the project. Management has set the team up so that it can make most of its own decisions—quickly. But a few decisions (covered at the end of this chapter and in greater detail in Chapter 10) must be made by top management. Because of its limited availability, it is important that upper management's decision points be kept to a minimum and that the essential decision points be carefully planned into the development schedule so as not to cause delay. It would be ironic to have top management cited as an excuse for delay after they had worked so carefully to close all the schedule's loopholes. Nothing demoralizes a charged-up team faster than a hurry-up-and-wait response from management. Although there may be many other equally urgent demands on management's time, the development team cannot see these, and any management delay will be interpreted by them as mere lack of interest.

Controlling a Project's Scope

Rapid product development comprises more than just rapid execution. As covered in detail in chapters 4 through 6, it is also essential to frame the product concept so that it is capable of being developed rapidly. Upper management has several opportunities to influence the way in which the product concept is framed, both initially and as the project evolves.

Initially, top management has an opportunity—indeed, the responsibility—to establish a broad product scope or global constraints on its design. For instance, an example in Chapter 6 described a manager who sensed a need to place a limit on the number of features to be considered in an electronic product that was threatening to get overly complex. He simply established the constraint that all the electronics would have to fit on one printed circuitboard of a certain size. Then the team could work within this limitation to make its own product feature trade-offs.

Without such limitations, newly available technologies and market pressures will encourage the product team to enhance the product concept continually in the phenomenon known as creeping elegance. This tendency destroys development schedules, because the team now finds itself aiming at a moving target. Top management's role here is to hold the schedule inviolate and thus control the creeping elegance.

As explained in chapters 4 and 12, the principle behind controlling a product's scope is to keep the product simple and get it out quickly, before market need shifts. If the product lags and the market shifts, the product becomes subject to redesign, which subjects the company to new risks of market shifts, creating a continual catch-up situation.

Although top management's role is to control this phenomenon through forcing adherence to schedule or global constraints on product scope, there will be legitimate cases when it is recognized that the product concept is seriously off the mark and it will have to be changed. This should be recognized as being a major upset, not just a deviation, and upper management should then renegotiate the entire project plan with the team. By making the restart special, management will help also to keep it rare.

Overseeing the Use of Resources

Even though our principle is to turn the vast majority of the decision making over to the development team and hold it accountable for its progress, top management still remains responsible for effectively employing the assets under the team's control. Consequently, upper management must monitor the team's progress to ensure that the company's assets are being put to good use. Chapter 10 explained how this could be done without delaying development or infringing on the team's tactical decision-making process.

Occasionally, changes will be needed in human, facility, or financial resource levels—usually in the form of an increase! When an increase or adjustment in resources is needed, top management must be prepared to respond quickly. This is one situation in which upper management should be keeping itself off the project's critical path, as discussed above. Doing so means two things here, first that top management should be informally keeping apprised of the team's progress so that the request for resources does not come as a surprise. Second, it means that top management should be planning its resources so that they can be made available on short notice if need be.

The overcommitment of engineering resources, which is a serious problem in industry, receives considerable attention in Chapter 11. Many products are late to market simply because of being starved for engineering resources, such as model-shop time. Once these bottlenecks have been discovered it is up to top management to bolster these resources so that time-critical projects are not left sitting in an obscure in basket in engineering while the market clock continues to tick.

Overcommitting resources is an enticing trap that is destructive to development schedules. The conscientious manager wants to ensure that his or her resources are being productively engaged, so it is natural to keep some development work in reserve to cover potential slack periods. Moreover, the ambitious leader normally expects a little more out of people than they can deliver. To cut development cycles, however, the challenge must be directed specifically at the schedule, not the workload. The point is to ask to get a product in half the time rather than to request twice as many products in hopes of forcing people to work faster.

On occasion a project will prove to be technically infeasible or else not viable as a business proposition. It is top management's responsibility to detect and analyze these situations and pull the plug on a project if necessary. If it is properly motivated, the team is operating in a charged-up entrepreneurial mode and may be oblivious to the storm clouds overhead. Stopping such a motivated group is one of the more difficult and unpleasant tasks of management. Usually, the best approach to aborting a project is to separate the project from the individuals (in an ongoing project the objective is to connect the individuals to the fate of the project) and offer them encouragement as you help them redirect their efforts. Fortunately, if the project has been set up and run according to the principles suggested throughout this book, this option should not have to be exercised very often. Nevertheless, it is still a key responsibility of management in its role of employing corporate assets constructively.

USING SHARP TOOLS

Sharp cutting tools, in the kitchen or the shop, allow work to be done faster, more effortlessly, and with greater precision. Used carelessly, they can also do more damage.

The same is true of some of the methods suggested in this book. They are sharp! Instead of accelerating product development, they can actually slow it down if they are used inappropriately. For example. the virtues of having more autonomous teams should be clear to anyone who has read up to this point. The underlying assumptions, however, are that the team's members want to be on the team (have in fact volunteered for it—see chapters 7 and 15), are mature, and are reasonably self-motivated. If these premises do not hold, team members may not know what to do or may simply choose to do nothing. If they are most familiar with having a strong functional boss to line up their work for them, they may not know how to start work on their own, particularly in an environment in which they have to ask for partial information (see Chapter 9).

Some organizations have lost all respect for schedules because management has cried wolf too often. In this case it may take extra management attention (or perhaps additional intermediate deadlines) to encourage team members to start working hard from the beginning of the project instead of waiting until the deadline is upon them, particularly if the individuals on the team are not accustomed to taking responsibility for schedules. Without the normal type of functional oversight and hard controls (see Chapter 10), a separate team could actually take longer to get the product to market than could a functional organization.

Chapter 4 recommended taking short, fast steps when innovating. Clearly, if an organization adopts the shorter-step philosophy without simultaneously quickening its pace considerably, it will move slower than before.

Finally, many employees are used to working under a heavy backlog. Chapter 11 suggested reducing project backlog substantially in order to cut queue time. This reduction cannot be taken as a signal to slack off, or else the benefits of faster cycle time will not appear.

These examples are intended not to dampen enthusiasm for trying these techniques but simply to warn that they are in fact sharp tools. Also, it should be clear from the references to other chapters that these techniques interact and should be considered a package. The best way to ensure that the techniques suggested here will accelerate rather than decelerate the development cycle is to first put them into place carefully on a small scale. Nurture a pilot project to ensure a success, and expand it from there. The next and final chapter provides guidance in making this transition into rapid product development.

SUGGESTED READING

Kiechel III, Walter. 1988. The politics of innovation. *Fortune* 117(8): 131–32. A clever lesson in the subtleties of mixed messages. Kiechel humorously illustrates how management unwittingly throws up roadblocks, such as the "no-special-treatment reflex."

Quinn, James Brian. 1985. Managing innovation: controlled chaos. *Harvard Business Review* 63(3): 73–84. Innovation, compared to other corporate activities, is a chaotic process to manage, accelerated innovation being even more so. In this classic Quinn provides an exceptionally lucid account of innovation and its management.

Schein, Edgar H. 1989. Reassessing the "divine rights" of managers. *Sloan Management Review* 30(2): 63–68. Schein stretches our thinking by suggesting that in the future hierarchical authority will probably play a much smaller role, while coordination skills will become increasingly dominant. One roadblock, he contends, is that we are virtually incapable of comprehending a nonhierarchical organization.

CHAPTER 15

Getting Started

The building blocks are now available to create a rapid product development cycle, but where do we start? One effective approach is to have a small, well-nurtured pilot development project, involving carefully selected and trained people, followed by cloning the techniques for use with other projects. This chapter explains why this approach works, and suggests what can be done to make it most effective.

The changes described in the preceding chapters are not just minor adjustments of procedures. In fact, to achieve the full potential of rapid product development demands a fundamental change in organization behavior. Experienced managers may have some scars demonstrating that it is much easier to talk about changing behavior than to do it. The change proposed here is indeed no less difficult than others the reader may have experienced.

Let us step back for a moment and recognize that all lasting changes in organization behavior ultimately require both the acquisition of new attitudes and a change in behavior. Psychologists have found that behavior and attitude travel together. It seems obvious that a change in attitude will normally result in a change of behavior. Nonetheless, attitudes can be inordinately difficult to change. Old attitudes are extraordinarily persistent.

This psychological reality applies directly to the business of developing products quickly. A large amount of training and education often results in only minor changes in behavior. This does not mean there is no role for education, but it is just not the star of the show. The path of education is a slow one; it simply does not produce change fast enough.

Fortunately, psychologists have discovered another interesting fact about attitude and behavior: if you change behavior, attitude will usually change along with it. The healthy mind works hard to bring its attitudes into alignment with its behavior, not just the reverse. For most people this is not intuitively obvious. It is nevertheless the key factor behind our approach to organization change.

JUST DO IT

There is value in trying to change attitudes, but we prefer to concentrate on changing behavior. This approach is more reliable, and far faster, than trying to attain a change in understanding that is sufficient to trigger an increase in organization capability. Try to get people to do things differently. Once they do, attitude change will follow.

This is not to say we avoid giving people the knowledge and the skills they need. Success requires an investment in skill building, which is valuable. A number of technical techniques have been discussed in this book, and education can help transfer these techniques. However, intellectual training pales in comparison with the urgency created by pushing somebody out of an airplane and telling them, "The ripcord is in front; you should probably pull it!"

AVOID THE IMMUNIZATION EFFECT

⚠ Though we strongly favor the path of taking action, not just any action will do. Setting up for a major success on the first try is imperative. In organizations that have attempted behavioral changes and failed, we have observed a heightened "immune response" to a reintroduction of the program. "We tried that once and it was a disaster" is the classic complaint. In many cases the initial attempt was poorly planned and executed, but the blame is typically placed on the idea, not its implementation. We call this phenomenon of heightened resistance to change after an abortive first attempt the immunization effect.

This phenomenon is strong enough that it should be carefully anticipated whenever attempting an organizational change. To use a biological analogy, to infect an organization with new ideas, be sure to use a strong enough dose of virus. If you do not, you will simply find it that much harder to reintroduce the idea on a second attempt.

In short, you should fight to win in organizational change. Better yet, don't try simply to win—try for unconditional surrender.

STACK THE DECK

To overpower a problem choose a small enough one, then attack it with more resources than necessary to solve it. Normally, this is done by creating a small-scale pilot program and doing everything possible to guarantee its success. This approach is quite different from the "controlled experiment" that tempts many managers. This traditional attitude says that we should try to pick a representative project for the change program, because if we pick an easy one and do many unusual things to ensure its success we cannot be sure the new techniques will really work.

The controlled experiment is a clinically attractive but ineffective approach. Our goal is to achieve a successful first experiment rather than a representative one. The successful experiment will build an organization's confidence and equip people with practical skills to take on more difficult projects. We are deliberately trying to stack the deck in favor of success. This project is not a scientific experiment but an attempt to build new organization skills. The key is to choose a solvable problem and overpower it.

SELECT THE RIGHT PILOT PROJECT

Because the primary objective of this project is to create a success, this will affect the criteria we use to select a pilot development project.

First, look for a project of moderate to short duration. Not everything will go correctly the first time, so we want a chance to do it again quickly. Just as incremental innovation can be used to manage risk in product innovation (see Chapter 4), short projects can be employed to reduce the risk of failure as the development process is modified. Look for a program that would take less than the average time to complete, even with the old methods. The shorter the first project is, the faster its lessons will become available for other projects.

Second, find a project large enough to be meaningful, but not one that creates a bet-your-company situation. If the task appears to be too easy, it may be viewed by the organization as trivial. If this happens, the project will get little management support, and its success will build little organization confidence and result in minimal learning. On the other hand, avoid excessively risky projects. Because you are already adding the risks of process change to those of product development, do not let these risks build to where you can't win or where management has to step in to rescue the team. If the project is too risky it will be difficult for the team to achieve self-management because the stakes will be so high that top management will feel it must be involved in too many of the project's decisions. So seek the middle ground of a meaningful project that is not too important.

Third, choose a program before either the product specification or its architecture are defined. The project must be structured for rapid development from the very beginning to have maximum impact on the development cycle (see chapters 4 through 6). To realize this impact, catch a project before it has started. Never select a project that is already in progress for a first attempt because the odds of its success are low. To do so loses control of key issues like product specification, team selection, system design, and so on. You may be lucky enough to succeed, but don't rely on luck. If you cannot control the front end of a project, don't even try to use it to change the organization—find another project.

The combined effect of these criteria can be quite restrictive. You may thus need to do some searching before finding the right project. Sometimes the answer will lie in selecting a relatively immature product idea that is in the planning stage, or in scaling back the scope of a larger project by using the incremental innovation approach discussed in Chapter 4.

GET THE RIGHT LEADER

Even before doing an economic analysis and creating the detailed product specification, select the project leader. As explained in Chapter 7, no other single decision has more influence on the chances of success. With the right team leader, most of the other steps that need to happen will take place. Selecting a good leader is like getting three aces in a poker hand on the first deal; it doesn't guarantee a win, but it makes the odds pleasantly high.

Let's be realistic, however. In most organizations not every person has the potential to lead a product development team. Because problems with capacity planning are likely occurring now (see Chapter 11), the odds are high that the person you want most is busy on sixty other things. It will be necessary either to free this person from those duties or find an adequate substitute. However, think twice before being convinced that only one person in the organization is capable of performing the task. Sometimes this means you are either excessively severe in evaluating your people or you are not looking widely enough for team leaders. Remember that you don't need a perfect team leader, just a good one.

Of course, the leader should be a volunteer, just as the team members should (see Chapter 7). Getting candidates to volunteer is actually quite simple. Be honest and explain the purpose of the project, its importance in establishing new attitudes, the high level of top management support and visibility it will have, and the demands it will likely place on the team. What's needed is not a sales job but rather an honest portrayal of the situation. Such an explanation should cause the type of individual sought to volunteer. If it does not, you have the wrong person anyway; find someone else.

Sometimes we are asked about what age and experience to seek in a team leader. Often younger, less-experienced leaders do a superb job, but we have also had equal luck with seasoned veterans. The veterans may take a little longer to acknowledge that many restrictions have really been removed, but they tend to be quite sure footed and pragmatic in their technical problem solving. You can find a winner in either group.

TEACH THE TEAM NEW SKILLS

Let us briefly remind you of some key points from earlier chapters. Economic analysis is a must. Also, a sound product specification must be developed. Then the correct team must be selected. The system architecture has to be properly structured, the team properly organized. Concurrent activities should be maximized, the right control systems established. Testing strategies need to focus on rapid learning. All these things are important and reinforce each other. They are the technical tools used to achieve organizational change.

These tools are a critical part of the change process because they give the organization skills to do things in a new way. To omit providing these skills may still develop a superb level of motivation and even a great deal of self-confidence, but such self-confidence would be unjustified. Imagine if the Navy tried to train fighter pilots simply by motivating them and building self-confidence. The cockpit of an F-14 would be pretty confusing to anyone who did not know what all the buttons and instruments were for. We need to balance concern for the "soft" issues with the need to equip people with the "hard" technical tools they need to do their jobs.

These tools equip a team to be self-managed so that decision making can occur at a low level. Yet tools alone are not enough; management must also provide the right leadership to nurture the new behavior.

LET THE TEAM RUN THE TEAM

As Chapter 14 stressed, the role of top management in rapid product development is less one of controlling than of facilitating progress. The inverted organization chart is an appropriate model for this approach. In it top management is on the bottom of the diagram supporting the efforts of the rest of the organization. This is an apt visual metaphor. The role of top management is to search for ways to help the team be successful, and vigilantly watch for problems that could slow its progress. Management's function is to enable and facilitate progress, not control it.

This facilitation is done through instituting new behavior patterns in the way projects are monitored and directed. As discussed earlier (Chapter 10), monitoring is best done daily or weekly, and informally. Because we want to place a minimal administrative burden on the team, we are biased toward informal verbal status reports, not formal written ones. Management by wandering around (MBWA) accomplishes this perfectly.

The role of management in directing a project can be easily summarized: let the team run the team. Only the full-time players have the information to make proper decisions. Give them both the responsibility and the authority to do this. Resist their probing attempts to delegate decisions back upstairs. Let them know they are responsible for their decisions, that they have the best information of anyone in the organization, and that you will rely upon them to make the best decisions possible.

Be careful to support the team in its decisions, even if this requires making a substantial change from historical behavior. Nothing saps initiative more than someone's second guessing the decisions made. This is doubly true when that person correctly spots mistakes. Such second guessing must be controlled, and the team must be supported so that it will develop its own new operating style.

Consider the true role of mistakes in the new setting. Minimizing mistakes is not the real objective. Granted that our approach will generate plenty of mistakes, because the possibility of mistakes is inherent in it. Deliberately trying to achieve a high level of concurrent effort causes us to rely on partial, often incorrect, information. We are trying to develop products quickly—not avoid mistakes. We are in fact trying to make a lot of mistakes quickly, not a few slowly. Mistakes are the signs that point us toward correct options. The team should thus not be crucified for making mistakes.

Nonetheless, there are times when it is proper for management to intervene with the team. First, management should do so when additional resources are required. This is the primary reason management is involved in monitoring the project anyway. When some other part of the organization is failing to support the team with money, time, attention, or other resources, management must intervene quickly to solve the problem. Rapid intervention is particularly crucial because it prevents the team from believing that their progress is being controlled by external factors. It is absolutely essential to success that the team feels the locus of control for the project to be within the team itself. Therefore, management must act quickly to deal with external factors that could slow progress.

The second type of management intervention is rarer. Occasionally the team will uncover information in the course of product development that suggests that the original goal is no longer achievable. In such cases management may have to intervene and modify the target. This should be done only with careful technical, market, and personal judgment, but sometimes it must be done.

This need occurs because we must start projects with imperfect technical and market information and sometimes we get surprised. Most of the new information is unlikely to invalidate the original approach because, as suggested earlier, of deliberately choosing a pilot project with low technical and market risk. Yet there are no guarantees.

In one project that we worked on involving the development of a multiuser computer system, management reassessed the feasibility of the original goal eleven months prior to its introduction. The project was under great time-to-market pressure. Having management eliminate some relatively minor, but difficult, software features cut the remaining development time to seven months. Such interventions can be critical to assuring the success of the team.

In addition to management's decision-making role it also plays a symbolic one, as suggested in Chapter 14. Ultimately, this becomes a question of whether top management considers the project important enough to devote a significant amount of their own time to it. Managing any business is a 120-ring circus with many activities competing for top management's attention. The whole organization watches carefully to see what gets the attention and uses these signals to assess what is really important to top management.

Senior managers must devote time to a project if they want anybody to believe they are seriously interested in it. No amount of memos or words speaks as loudly as how we allocate our time. People believe what we do, not what we say.

The most effective way to spend this time is to spread it throughout a program in tiny chunks. Avoid the more typical "binging" behavior of spending large chunks of time at very distant intervals. Frequent short contacts allow the detection of problems and provide added support earlier in the development process. They also reduce the time between the origin and the resolution of problems. Furthermore, frequent contact maximizes the impact of the symbolic communication on the importance of the project.

SEARCH THE EXPERIENCE BASE

While making organizational changes it is important to realize that there are secret allies in the organization. Every organization has much more raw material for making organizational changes than it usually realizes. Hidden within the experience base of every organization are past actions that will closely resemble many of the techniques we have been describing. For one reason or another these actions did not blossom into company-wide approaches. Yet these past acts can be an extraordinarily valuable resource if used properly.

Employing past actions can do two things for us in the change process. First, it gives examples of behavior that is directionally correct. These examples can then be reinforced to achieve the targeted behaviors. Such concrete examples are meaningful to an organization.

"Remember the time we built *two* prototypes of model X? Well, maybe we should do it again."

Second, these past actions validate that business can be done differently, even in one's own organization. This point is critical for the change process because the organization must believe it can change before it can gather true enthusiasm for the change process. Simply showing the desirability of change is not enough, because there will still be fear of the consequences of change and the loss of control that can accompany it. Showing real examples of similar actions that have been taken in the past without suffering disastrous consequences can reduce some of the fear that comes with change. Reducing this fear and building on the seeds of past actions, however small they may be, can accelerate the change process.

CLONING THE PROCESS

When a pilot project is nearly complete, begin planning the next few projects to make use of its techniques. We call this "cloning the process." Generally, a gradual roll-out is better than an explosive one. In other words, do not get carried away initially and try to scale up too quickly. Rather, continue nurturing subsequent projects with ample attention and resources until a pool of experienced people is built. It is better to do one project, then two more, then three more, than to do one project, then five, then twenty-five. The latter approach simply dilutes management attention excessively, increasingly the likelihood of failure.

Also, the most viable way to pass the virus of rapid product development is through direct contact, not simply by exposing people to the written word. Having people work on teams where development is done quickly lets them acquire the key "soft" values that are the core of the techniques. Let them see what works and what doesn't. They will unconsciously be recording a set of behavioral options that they will be able to draw upon when they discover themselves in similar situations. There is just no substitute for on-the-job training in these techniques.

Furthermore, shoulder-to-shoulder contact develops confidence in an approach that cannot be generated simply by reading about it. Having this confidence is crucial, because in the bumpy ride toward a product introduction there will be temptations to slow down when things are not going well. Confidence in the approach will maintain momentum. This confidence is best built through participation in successful projects.

WHEN DO WE STOP?

Once the cloning process has started, should it take over the entire organization? It would be a disservice not to warn again against the illusory nature of universal solutions. We must reemphasize that many development programs should not have development speed as their primary emphasis. An organization that cannot emphasize the other objectives described in Chapter 2, such as product performance or product cost, when these objectives are appropriate will simply have exchanged one set of weaknesses for another.

Reviewing Chapter 2 will help you determine when a certain objective is appropriate. We have found no universal answer other than doing the analysis and wrapping the problem in numbers. Doing so is the best way to determine the development approach most appropriate for a particular situation.

Most organizations have to acquire the ability to do development in different ways to meet different objectives. The best organizations have alternate tracks that a development program can be put on, depending on the needs of a particular situation. Rather than saying "We do development in this one way," they say, "Tell us the product development problem and we will tell you which way of doing product development we feel would be most appropriate."

Even if rapid development were judged appropriate for all projects, there are still practical limits to the number of rapid development projects that can be handled simultaneously. These projects will demand certain scarce resources, such as qualified volunteers and top management's attention. Each organization must discover its own capacity to do rapid development, recognizing that when everything is a priority, nothing is a priority (see Chapter 11).

All organizations have their own great skills in doing certain things, and these skills should not be obliterated while building new capabilities. Try to preserve old, valuable skills for use in other situations. Remember that we are trying to create a multimode development organization, not one that can develop products in only one way.

There are two answers to when to stop. First, stop before destroying all the old skills. They can be quite useful in a complex world. Second, with regard to when to stop trying to improve the process by making it more capable and flexible, the answer is obvious: never.

THE USE OF CONSULTANTS

We are sometimes asked about the role consultants can play in reducing development cycles. It should be clear from our very existence as consultants that certain companies feel they can accomplish their objectives better by using consultants experienced in making these changes than they can on their own. Let us take a moment to examine this logic.

There is no particular magic to rapid product development. Any organization that carefully examines its own behavior and diagnoses the causes of its successes and failures is capable of developing the techniques and skills needed for quick development. It takes time and effort, but not magic. The issue is not whether an organization can develop these skills, but how long it will take to do so. As with any other major process of organizational change, such as JIT and TQM (total quality management), such a capability can be developed faster by building on the experience of others than by developing all the techniques oneself.

Just as one can buy time in the development process (see Chapter 11), companies that use consultants to accelerate their development process use them primarily to buy time in their change processes. They know they could do the job themselves eventually, but eventually is not soon enough. They want to get maximum value from their efforts and avoid going down blind alleys. They treat the business decision to buy time in the change process in the same businesslike way that they no doubt decided to buy time in their development schedule.

The approach to take should be familiar by now. Determine the value of a more rapid development cycle, then estimate how long it would take to develop this capability on your own. Then estimate how much earlier it would be possible to achieve this capability by using outside assistance. The financial value of shifting this capability forward in time then gives a rough measure of the value of outside assistance.

For example, if a business with $50 million in sales could improve its profit margins by one percentage point with a faster development capability, profits would be higher by $500,000 each year. If this increase could then be achieved six months earlier by using outsiders, the value of the consulting if it is 100 percent successful would be $250,000. Looked at another way, if the outside assistance cost $50,000, it would have to have a 20 percent or better chance of accelerating the change by six months to justify investing in it.

The point is that any decision to use outsiders should be based on whether they are a cost-effective way to achieve an organization's business objectives. The decision should be driven by whether or not they can buy time and whether this time has sufficient financial value to justify the investment.

Companies that use consultants feel they are jumping off from a higher starting point. They want access to a broad range of the options that can be used to accelerate product development and to know what the potential pitfalls are so that they don't have to invent every mistake on their own. Often they also use consultants as a tool for selling the organization change within the organization.

Anyone choosing to use a consultant should recognize that they will be most valuable at the very beginning of the change process, before broad organizational enthusiasm has been generated. During this fragile period of the change process you will want to stack the deck to ensure being as successful as possible. Later, as the techniques take hold and the process is cloned, you will generate a larger group of people infected with the new technique. Under these conditions consultants should have a minimal role, perhaps simply keeping you informed on the state of the art in this key skill area.

GET GOING

As we send you out to change management processes, we should point out that the changes we have described are additive. Benefits will derive from any of the techniques we have described, if they are properly implemented. Because they are mutually reinforcing, they work best when used as an integrated set, but do not let roadblocks in one particular area prevent you from making changes in others. Any successful step in roughly the right direction will help.

The ability to develop products quickly can be a tremendous competitive weapon. Start making changes now, and strive for continued improvement beyond any apparent limits. This is a new field dealing with a complex problem. Not all the answers have been discovered yet, and you are sure to discover some of them. Good luck; do start now.

We are reminded of the story of the wise man who was approached by a youth who asked him how long it would take to walk to Athens. The wise man simply replied, "Get going!" The youth, disappointed, concluded that he would not get a straight answer and began walking in the direction of Athens. After the youth had taken about twenty steps, the wise man called out, "About four hours!" The youth turned and said, "Why didn't you tell me that when I asked before?"

"First I had to see how fast you walked," said the wise man.

SUGGESTED READING

Schaffer, Robert H. 1988. *The Breakthrough Strategy*. Cambridge, Mass.: Ballinger Publishing Co. This book, a bit misleadingly titled, has nothing to do with building strategies but rather with getting things done in an organization through a sequence of small wins. These wins build confidence and ultimately develop the organizational capability to operate more effectively. A product development project may be a large task for Schaffer's small-wins tactic, but many of his ideas apply nevertheless.

CHAPTER 16

Assembling a Rapid Development Process

In the years since this book was first printed, many product developers in countless companies have tried these tools. We have worked with many companies to implant the techniques of rapid product development in their organizations. Our readers have learned, our clients have learned, and so have we. This chapter presents key observations gleaned from working with firms in many industries as they used this book in their organizations.

While other chapters of this book are designed to be read independently, this chapter integrates information from the previous chapters. Some familiarity with the rest of this book will enhance the value of this material.

MOST FRUITFUL AREAS FOR IMPROVEMENT

Although each company applies the tools in this book differently, some tools are clearly more popular than others. Clients are emphasizing:

Co-location (Chapter 8). When we present these tools, this one prompts the liveliest discussion. People who have actually participated in a co-located product development team including product engineering, marketing, and manufacturing, all have found it to be an effective tool and would choose it again if they wanted to get a product to market quickly. They are impressed with the ease and speed of communication and decision making. Those who have not experienced co-location themselves have different reactions. Some clearly see its assets, while others argue that it has no net positive effect. There are strong personal concerns about lack of a "home" or status, lack of privacy, or distractions. Opposition seems to be highest among software engineers, who often do need long uninterrupted periods to be productive, and in high-tech industries where the use of modern electronic means of communication have, to a degree, replaced face-to-face conversation. Decentralized or global operations also present a practical difficulty in implementation.

We weigh heavily the experience of those who have actually tried the technique and our own experience in watching it work in many settings. Thus, we believe the power of co-location is so great that it cannot be dismissed by anyone setting up a team. For example, we all have experienced the delay inherent in phone tag and its e-mail equivalent. Regardless of the progress of electronic communications in recent years, there is still a great deal more "bandwidth" in a spontaneous, in-person conversation.

Project overload (Chapter 11). This is an opportunity that often allows cycle time to be cut by half or more. The approach is to load as much staff on a project as it can support effectively. Companies that are effective at it initially cut their project list by 50 to 90 percent and are religious about not starting a new project until it can be fully staffed. They always ask, "Where could I put one more person on this project and reduce cycle time commensurately?" They keep loading until they reach saturation. As simple as this sounds, it is not easy, it requires a common understanding about project loading throughout the organization and a great deal of discipline to say no.

Although this tool can be used effectively alone, often it is a prerequisite for undertaking other actions. For example, how are you going to staff a pilot project or dedicate resources to a project if everyone is already booked at 160 percent?

Pilot projects (Chapter 15). Many firms recognize that it is not just one item that needs improvement but a whole set of values built into their culture that mitigate against speed. Consequently, they use a pilot project as a model of the new values and styles they wish to implant. This is a powerful tool that has helped many organizations make the transition to rapid development quickly and reduce development times to one half or less of their former value.

A pilot project's main shortcoming is that it is so attractive that management forgets to address the fundamental changes in approach that must be adopted to create enduring change. Some companies jump into a pilot—or worse yet, pilots—with little awareness of their weaknesses or objectives in accelerating development. One product gets to market quickly, but nothing is learned; later projects revert to the old system. Other companies climb on the pilot project bandwagon to get a pet product idea to market quickly. Because they put the product ahead of the process, the process isn't replicated.

Dedicated resources (Chapter 7). This tool is used as a starting point by many of our clients. First they assign product engineers and designers full-time to one project, then they add a full-time manufacturing engineer and progress to the level where they can support technicians, marketing people, test engineers and others, as the project's nature dictates. For some projects and cultures this is easy, and for others it is a real test of management resolve. Often it highlights a chronic project overload condition that must be addressed first. There are several benefits: concentrated staffing naturally cuts cycle time; time wasted switching between projects is eliminated; accountability for results and schedule compliance is much better; and dedication sets the stage for co-location.

Fuzzy front end (Chapter 3). Almost everyone seems to recognize that the portion of a development project before staffing is assigned represents a huge and inexpensive opportunity for improvement. Typically, at least three-fourths of this cycle time can be eliminated. The most popular means of dealing with this opportunity is to establish a process for handling new product ideas. Such processes tend to be successful if they include clear selection criteria and staffing assigned to work the process. If either of these is missing, the compressed process falls into disuse.

More ambitious firms often recognize that they also need to develop key technology offline. This step is more difficult because it requires careful technology planning and setting aside some current development resources to fill the technology storehouse. Finally, some firms refuse to deal with the front end because they have conveniently defined the project to start when the concept is approved; we can only hope that their competitors have chosen to play by the same rules.

Better product specifications (Chapter 5). This improvement often starts with a particular multifunctional or pilot team, rather than from a comprehensive redesign of the specification process. It is often supported by field-based market research performed by the team. The result is a more solid understanding of customer requirements and an ability to adjust the spec as needed according to new information arising during development, which is quite different than the notion of designing to a "frozen" spec. For example, one team in a medical electronics company that prided itself on miniaturization developed a small, pocket-sized version of their desktop instrument. They were proud of their accomplishment, but when they took a prototype to some hospitals, customers told them they were concerned about theft of such a small package. So the team adjusted quickly, making it bigger. Such advances in specifications usually happen when the marketing person on the team puts in extra time on the project.

Clearer strategies. As markets expand, technologies settle out, or operations become more global, firms often become unsure of who they are serving, why, or how. Without a strategic foundation, development groups are subject to constantly changing opinions, which means rework and ultimately a loss of the sense of urgency. Some companies have responded by creating business and product strategies and communicating them to development groups. Dissemination of strategies tends to be more of an issue in privately held companies where strategies may be closely held by the owner.

Weak linkages. For product development to work effectively, at a minimum we need strong linkages between marketing and engineering and between engineering and manufacturing. Usually one of these links is clearly weaker than the other, often stemming from deep-seated cultural or organizational structure issues. Many companies are identifying and strengthening the weak link. When the marketing link is weak, projects often experience infrequent but major redirections or are in constant flux. When the weakness is on the manufacturing side, typically there are many surprises—and delays—as the project moves into manufacturing. Solutions depend on the source of the weakness and may include organizational changes, hiring, training, or co-location. Often, we may make a fast start on the solution by bridging the weak link with a dedicated team member.

LEARN BY DOING

By now, many companies have undergone some type of total quality management program. Although some aspects of such programs may tend to slow product development due to the extra requirements and paperwork involved (see Appendix B), we find that companies that have worked through total quality are often in a stronger position to streamline their development process. They have learned some basic skills, such as conducting meetings and root-cause analysis. They tend to think more multifunctionally and with more of a process orientation. They know how to construct metrics to measure progress. Most importantly, they appreciate that real progress will take time and require a substantial commitment of effort, which tends to come at the expense of more immediate payoffs.

To avoid diverting line managers from their immediate duties, some firms establish a staff function, such as a quality function, to absorb the extra burden of improving how business is conducted. There would seem to be value in having a staff specialist lead such changes. Yet, we find that the staff specialist has difficulty in getting line managers to buy in and take the program seriously. Often the staff managers create plans, processes, or paperwork that are overly abstract or complex. If fast cycle time is important to your company, then the chances of success are much greater and progress is likely to be faster if key product development line managers take the lead in making the change.

Don't Depend on Osmosis

Thousands of copies of this book now sit on product development managers' shelves. Purchasing the book is the easiest step on the road to rapid product development—and unfortunately the last one for some individuals.

Rather than just buying copies for your employees, consider taking a more active role. Many people have learned a great deal from this book by starting a study group. Each week a group member leads the group through a chapter, considering how it could be applied to their business.

In some organizations an offsite meeting has a better chance of capturing the attention of key participants in the development process. The key to such meetings is that they translate concepts into action, at the end forming action plans with responsible individuals and due dates. This book then becomes a source of ideas and a reference;

organizational change comes from the action of individuals, not from osmosis.

Learning from Projects

The most powerful way to assimilate these techniques is to try them out on actual development projects. Although companies do get marketable products out of such projects, they seldom pursue the second benefit of product development, which is to consistently learn from projects. Thus, the same flaws are repeated on each project, and improvement is slow. Our leading clients are reviewing in-process and completed projects regularly to capture the lessons learned so they can modify their development processes to institutionalize this knowledge. Of course, this takes some time away from actual development, but these companies regard the time spent as a long-term investment in competitive capability.

PRODUCT DEVELOPMENT AS A PROCESS

Since this book was originally published, managers have started viewing their businesses more as processes, and many companies have revamped their product development processes. Each firm takes its own approach. One is reengineering, which concentrates on value-added analysis and opportunities for computerization. Another is stage-gate systems, a refinement of phased project planning (see Chapter 9). And another is "product life cycle" systems, which includes supporting the product once it is developed.

Just as product development is more than solely an engineering responsibility, it is also more than just a process. A purely process orientation misses organizational structure opportunities such as colocation. Nor does the process capture the fundamental tools of project economics or capacity management. The process is a portion of product development, but it is not an adequate overall framework.

When assessing your process, it is important to keep in mind what should be driving it. Such objectives are often implicit or hidden. If you wish the process to facilitate rapid development, speed must be an explicit objective. Otherwise, more traditional objectives, such as conservation of material and human resources, will implicitly drive your development process. Project economics can help you set the proper balance.

Many development process designs implicitly conserve management's time by installing several gates or checkpoints in the process. But management time can be traded for cycle time by using fewer gates and more MBWA (management by wandering around). How far dare you go with this approach? Well, one of our clients, a global producer of motor vehicles, develops new platforms with only two gates, a starting one at concept approval, and a finish gate when they ship the first production vehicle.

THE TOOLS REINFORCE EACH OTHER

In writing this book we followed our own advice from Chapter 6: break the subject of rapid development into chapters which are manageable, coherent modules. As explained in Chapter 6, this modular approach has solid advantages in limiting complexity, making it easier to comprehend the material both while writing it and when reading it. We provide cross references to related material in other chapters, but one could get the impression that the individual tools stand alone.

This seeming independence is misleading. Our experience in helping clients to implement these tools over the past few years reinforces the principle that the real power of these techniques lies in their synergy. The whole tool kit is indeed more powerful than the individual tools used alone, due to reinforcing effects. Here we describe three underlying themes that tie the tools together.

The Cost of Delay

Some readers have skipped right over Chapter 2 (Wrapping It in Numbers), because it seemed a bit boring. Apparently, they were eager to get to the heart of the book. This tool *is* the heart of the book, for two reasons. One is that clients and participants in our workshops who have calculated and used project economics generally have made the fastest progress toward rapid development. Second, we have discovered that this tool provides the motivation and underpinning for everything we do in reducing cycle time. Because we are in profit-making enterprises, we really have no business working on cycle time until we are sure that it provides profitability benefits greater than alternative investments of money and energy.

Too many managers grab onto a cycle-time program because of its popularity rather than looking at it as they would an investment in a new manufacturing plant. In fact, rapid development may not apply well to their business or to all of their products, and in any case, these fad-chasing managers quickly lose interest in the program. Although it is our purpose in this book to dwell on time-cutting tools, we advise that they be applied, not blindly, but as their costs and benefits dictate in a particular situation.

Complexity Drives Schedules

As explained in Chapter 4, and illustrated in Figures 4-2 and 4-3, complexity translates into time, but the connection is often hidden, thus overlooked. Moreover, complexity appears in many other contexts than just the incremental development covered in Chapter 4. For example, in Chapter 6 we describe the power of a modular architecture, not in its traditional technical role, but as a managerial tool to shorten development cycles. Underlying this tool is the notion that a modular structure manages complexity by largely containing it in comprehensible amounts within modules. For instance, software developers often apply the modular architecture tool by breaking complexity into two pieces—intermodule coupling and intramodule cohesion—then setting objectives of minimizing coupling and maximizing cohesion. The same principles can be useful in hardware development. Focusing on complexity helps us to set up an architecture that makes best use of the schedule time available.

In large projects it is difficult to hold to our recommended team size limit of about ten people (see Chapter 7), but modular approaches allow dividing the project into many teams of ten or fewer. Again, this is an issue of managing complexity in staffing a project: with large teams, coherence becomes an issue, but with more numerous smaller teams coupling gets out of hand. An alternative to this team setup dilemma is to undertake a less ambitious project (Chapter 4), which is another means of employing complexity as an ally.

Teams Make the Other Tools Work

Development teams are another unifying theme that enable many of the other tools to be effective. For example, we explain in Chapter 5 the value in "owning" the project that comes from having a multifunctional development team write its own product specification, thus establishing an important linkage between specification writing and the team material in Chapters 7 and 8. This strong team is then in a much better position to deal with specification changes as they arise, because the team has all of the information internally to make a fast, accurate decision on the impact of a specification change. As obvious as this may seem, it runs counter to the advice that one can only get to market quickly by "freezing" the specification which has frequently appeared in the literature on rapid product development. Although engineers may prefer frozen specs, such specs do not fit with marketplace realities. Consequently, we believe that a flexible but fast solution is to hold the team jointly responsible for specification changes. Clearly, this requires a real multifunctional team, not just a group of departmental representatives acting under the banner of a team.

Many companies have discovered the value of getting real manufacturing involvement in early product design issues, a topic covered in Chapter 13. Given that most manufacturing people are governed by short-term crises, one of the most powerful approaches we have found to get this quality time when it will be effective is to assign a manufacturing engineer to a development team full-time from day one. Companies that are not willing to take this step usually find that they are unable to break the ongoing pattern of late manufacturing involvement and tail-end delays as manufacturability issues are resolved on the factory floor. Thus, teams enable the transition to real concurrent engineering.

HOW DO WE PROCEED?

We wish we could give you a standard recipe for speeding up your development process, but it is not quite that simple. To begin with, we assume that you have already taken the obvious, inexpensive steps; what lies ahead is bound to be more complex.

Some authors suggest a certain sequence, which involves simplifying and eliminating unessential steps first, then applying overlapping while compressing delays, and finally supercharging what is left. Our model is to look for the biggest opportunity currently, in a cost-benefit sense, then move on to the next opportunity.

This fits well with the total quality concept of continuous improvement. It also fits with just-in-time manufacturing's rocks-and-water metaphor, in which inventory levels—in our case, cycle-time

TABLE 16—1

Troubleshooting Guide

Symptom	Action
One or more specific bottlenecks for all projects	Fix these bottlenecks first using process redesign, technology solutions, or more resources (justified through value of time analysis)
Differing opinions on time to market emphasis	Calculate value of time (Chapter 2)
Upper management lacks interest in time to market	Build a case for speed by studying lost market opportunities, benchmarking competitors, or calculating the value of time; be alert to more basic impediments, such as project overload or management turnover
Employees are unsure about committing to a time to market program	Conduct cycle-time training, allowing for plenty of discussion of the issues and for developing actionable grass-roots plans
Constant flux in staffing and priorities; projects staffed thinly	Analyze development capacity; if insufficient for the projects underway, kill some projects (Chapter 11)
Time vanishes in meetings, phone tag, and e-mail backlogs	Co-locate (Chapter 8)
Cultural issues, such as risk aversion, overemphasis on cost, not invented here, or chronic schedule slippage	Initiate a pilot project (Chapter 15), preceded by adequate analysis, so that the pilot can overcome the specific impediments involved
Shifting project requirements, project redirection, creeping elegance	Consider marketing's role which may be weak, intermittent, or dictatorial
Tail end surprises in manufacturing startup	Analyze manufacturing's role in influencing design decisions early in the cycle (just attending meetings does not count)
Inability to make firm decisions on project direction	Develop and communicate product line and technology strategies, backed up by a focused business strategy

levels—correspond to the water level in a stream. As we lower the water level, rocks on the bottom start to appear. Our job is to identify and remove these rocks, so we can lower the water further. The process is continually repeated. One advantage of this small-steps approach is that we begin to see the fruits of our work sooner, which encourages further improvement.

Some clients complain that the problems never go away, they just come back again later. This is true; there are more rocks under the ones we remove, but they are different, deeper rocks. It is like going up a spiral staircase: the same view reappears a bit later, but we have ascended in the meantime. When clients measure their progress, the gain in elevation is apparent even though the same issues reappear.

Although the recipe does vary, we have observed some general connections between certain symptoms and likely actions to resolve the underlying causes. Table 16-1 provides the most common connections at the highest level. This short "troubleshooting" guide may help you locate the most fruitful places to get started.

Like rocks and water, the improvement process, in general terms, involves doing some analysis, identifying the most important current issues, then taking action on just one or two of these. This is completely different than redesigning the whole process, assuming that everything is equally broken.

Best wishes on your journey! We will be watching for you on the spiral staircase.

REMINDER LIST

This list summarizes key recommendations presented in the first fifteen chapters. If you are not familiar with the rest of the book, the list can create the comfortable illusion of understanding the concepts. Please read the book before using the list.

We strongly recommend that you do not use the list as a checklist or "report card." Used in this manner, these cryptic reminders provide no real assessment of your level of proficiency. For example, consider the first reminder. Virtually all companies do some economic calculations in conjunction with development programs, but few use time-sensitive models such as those described in Chapter 2. In addition, "scoring" yourselves on the whole list is likely to dilute your attention over too broad a set of issues. You will be more successful if you do a few things on this list superbly rather than a little of everything.

Chapter 2: Wrapping It in Numbers

1. Do we create economic models for development projects? p. 17
2. Do we quantify the value of the four key development objectives? p. 20
3. Are development tradeoff decisions based on economics? p. 20
4. Does the entire team understand the model? p. 28
5. Is the modeling done by a cross-functional team? p. 29
6. Do we calculate tradeoff rules for projects? p. 38
7. Do we spend extra time on the pricing and unit volume assumptions? p. 40

Chapter 3: The Fuzzy Front End

1. Do we have objectives, schedules, and budgets for front-end activities? p. 43
2. Do we know the cost of delay during this stage? p. 44
3. Do we assign a clear responsibility for this stage? p. 46
4. Is there a system to manage the dormant projects list? p. 47
5. Do we measure time with a "market clock"? p. 49
6. Does the business planning process delay project start? p. 51
7. Does a new products committee meet often enough? p. 51
8. Does an ad hoc committee have adequate resources? p. 51
9. Are handoffs from a new products group effective? p. 52
10. Do we limit initial planning to critical issues, saving other issues for later? p. 53
11. Do we have a quick-react plan for an unexpected new product idea? p. 57
12. Is technology development done off the critical path? p. 59
13. Is there a clear long-term product plan known to the entire organization? p. 59
14. Are product plans linked to business strategy? p. 59
15. Do people move with the product from the front end into development? p. 60

Chapter 4: Innovating Incrementally

1. Do we treat new products as a tool for tracking market trends? p. 70
2. Have we picked a clear freeze point for the specification? p. 76

3. Do we have a clear product strategy on a product line basis? p. 77
4. Is the product strategy communicated to everyone involved in the project? p. 77
5. Do we view cannibalization as an opportunity to beat the competition? p. 78

Chapter 5: Product Specifications

1. Are specifications viewed as inherently incomplete? p. 85
2. Is a single cross-functional specification document used? p. 87
3. Does the specification identify the product's key competitive advantage? p. 88
4. Are customers involved in helping prepare the specification? p. 88
5. Are the needs of users and non-users considered? p. 90
6. Can we provide the phone number of a representative target customer? p. 90
7. Does the specification keep design options open? p. 91
8. Are all business functions involved in writing the specification? p. 92
9. Is the specification **jointly** developed by all business functions? p. 92
10. Is experience from past products captured in checklists? p. 93
11. Is brevity viewed as sign of a good specification p. 94
12. Is the specification developed in a face-to-face workshop? p. 94
13. Are we willing to start development with a partial specification? p. 97

Chapter 6: The Subtle Role of Product Architecture

1. Do we use modular architectures? p. 101
2. Are rapidly changing subsystems separate modules? p. 102
3. Do we minimize the movement of functionality between modules? p. 103
4. Do we have adequate performance margins on all subsystems? p. 103
5. Are interfaces between modules stable throughout the design process? p. 104
6. Are interfaces between modules robust? p. 105

7. Do we use standard interfaces whenever possible? p. 105
8. Is the system partitioned to simplify interfaces? p. 106
9. Is risky technology concentrated in a few modules? p. 107
10. Do we have contingency plans for risky modules? p. 108

Chapter 7: Staffing and Motivating the Team

1. Are there ten or fewer members per team or subteam? p. 111
2. Are team members volunteers? p. 111
3. Do team members serve from the beginning to end of program? p. 111
4. Are team members assigned full time? p. 112
5. Do team members report solely to the team leader? p. 112
6. Are all key functions represented on the team? p. 112
7. Are team members located within conversational distance of each other? p. 112
8. Has the choice of leader been publicly communicated? p. 112
9. Does the team have a powerful external sponsor? p. 114
10. Does someone on the team act as a technological gatekeeper? p. 114
11. Does the team leader think cross-functionally? p. 115
12. Does the team leader know how the company really works? p. 115
13. Does the team leader have good people skills? p. 115
14. Does the team leader have a clear vision of the product? p. 115
15. Do we choose the team leader based on skills, not functional affiliation? p. 118
16. Do we have enough generalists on the team? p. 121
17. Are team members willing to work outside of their specialty? p. 122
18. Do team members have equal status? p. 122
19. Are suppliers considered as possible team members? p. 124
20. Are suppliers deeply involved early in the program? p. 124
21. Are rewards based on group performance? p. 128
22. Are rewards visible to others in the organization? p. 129
23. Do we publicly recognize successful teams? p. 129
24. Are the sacrifices of spouses and families recognized? p. 129
25. Are follow-on opportunities planned for team members? p. 130

Chapter 8: Organizing the Team for Action

1. Does the team have the authority to make most project decisions? p. 134
2. If functionally organized, are resources for critical projects protected? p. 136
3. If a lightweight team, are decisions outside team timely? p. 136
4. If a balanced organization, are clear boundaries defined? p. 137
5. If a heavyweight team, are functional interests adequately represented? p. 138
6. If an independent team, is administrative overhead shifted elsewhere? p. 140
7. If an independent team, is integration with primary business addressed? p. 140
8. Is a viable alternative to primary organizational approach maintained? p. 140
9. Is potential for animosity between teams and functions managed? p. 141
10. Do team leaders think they have been delegated enough authority? p. 141
11. Are fringe player situations avoided if possible? p. 142
12. Do fringe players participate on a well-defined, scheduled basis? p. 142
13. Do external experts have well-defined work packages and milestones? p. 142
14. Are team members involved from beginning to end of the project? p. 143
15. Are team members capable of playing more than one role? p. 144
16. Are handoffs in the process minimized? p. 144
17. Does the team leader report to the general manager? p. 145
18. Is team co-located? p. 145
19. Does team workspace use "wall-less" layout? p. 146
20. Is team isolated from external distractions? p. 146
21. Are key team members located within listening distance of each other? p. 148
22. Does team approve the location and layout of the team work area? p. 150

Chapter 9: Achieving Overlapping Activities

1. Can we start procurement with an informal drawing? p. 160
2. Can we do initial production with interim tooling? p. 160
3. Can we use defective first-article parts to test downstream processes? p. 160
4. Are sequential approvals of engineering changes avoided? p. 161
5. Do team members feel they own the schedule? p. 163
6. Are outsiders used to challenge schedule assumptions? p. 163
7. Are schedules built by working from ship date backwards? p. 163
8. Do we learn scheduling tricks from other projects? p. 164
9. Do downstream activities pull information from upstream ones? p. 165

Chapter 10: Monitoring and Controlling Progress

1. Are there special control systems to facilitate speed? p. 172
2. Are written reports to management kept brief? p. 175
3. Does management visit regularly on-site with the team? p. 176
4. Does team have regular daily or weekly meetings? p. 176
5. Does team have regular conference calls with remote sites? p. 177
6. Do team meetings focus on future actions vs. reports of actions taken? p. 177
7. Are control systems adjusted to the size and phase of the project? p. 179
8. Is there a codified development process? p. 180
9. Are management decision points clearly defined? p. 181
10. Is the critical path of the project well-defined and understood? p. 181
11. Does the project have frequent measurable goals? p. 183
12. Are project's intermediate goals assigned to individuals? p. 183
13. Is project schedule posted in the team room? p. 184
14. Are technical and business reviews kept separate? p. 185
15. Are informal project reviews held in the team work area? p. 186

Chapter 11: Capacity Planning and Resource Allocation

1. Is there an adequate budget for travel to customers and suppliers? p. 190
2. Do we encourage using cheap, simple models? p. 191
3. Do we have the capacity to build models quickly? p. 191
4. Do we err on the side of building extra prototypes? p. 192
5. Do we consider the cost of delay when evaluating outside resources? p. 192
6. Do we consider partial and soft tooling for initial production? p. 193
7. Will we risk reworking tools to gain cycle time? p. 193
8. Are we constantly pruning the dormant projects list? p. 195
9. Do we refuse to start a project unless full staffing is available? p. 196
10. Do we calculate the financial impact of overloading our capacity? p. 196
11. Do we concentrate more resources on fewer projects? p. 200
12. Do we provide excess capacity for unanticipated projects? p. 202
13. Do we have a priority list for our projects? p. 202

Chapter 12: Managing Risk

1. Do we use checklists to avoid repeating technical problems? p. 209
2. Do we use market research to reduce market risk? p. 211
3. Do we use close customer contact to reduce market risk? p. 211
4. Do we have customers on the development team? p. 211
5. Do we get market feedback on our concepts and prototypes quickly? p. 212
6. Do we keep the design flexible on unresolved issues? p. 212
7. Do we keep high risk areas visible? p. 216
8. Do we have early warning indicators in high risk areas? p. 216
9. Do we independently resolve market and technical risk? p. 216
10. Do we have back-up plans for critical items? p. 218
11. Do we focus on quick, simple tests? p. 219
12. Do we combine testing and analytical methods effectively? p. 219
13. Do we focus tests on critical features first? p. 220
14. Do we first test at the lowest possible subsystem level? p. 221

Chapter 13: The Product–Process Duo

1. Do configuration controls vary with the stage of the design process? p. 227
2. Are large batch transfers of engineering drawings avoided? p. 331
3. Is key process engineering expertise applied early? p. 232
4. Do process engineers get experience in production? p. 234
5. Do we encourage the reuse of parts? p. 235
6. Do design engineers have access to process capability data? p. 235
7. Do we change product and process at different times? p. 237

Chapter 14: The Role of Top Management

1. Does top management have a rapid development objective? p. 243
2. Is this objective broadly communicated throughout the organization? p. 243
3. Is the rationale for this objective broadly communicated? p. 243
4. Do team leaders alert management to special accomplishments by team? p. 246
5. Do we realize the importance of making mistakes? p. 247
6. Is top management time focused on the front end of projects? p. 248
7. Does senior management help select the team leader? p. 249
8. Is the team allowed to form its own identity? p. 249
9. Are external reasons for schedule slippage eliminated? p. 249
10. Is there an implicit/explicit contract between the team and management? p. 250
11. Does the team leader report to general management? p. 251
12. Does top management reinforce the team's authority? p. 251
13. Does top management avoid delaying project decisions? p. 252
14. Is the team encouraged to make most project decisions? p. 252
15. Is top management available to the team on short notice? p. 253

Chapter 15: Getting Started

1. Do we focus on changing behavior rather than attitudes? p. 258
2. Do we stack the deck to guarantee success on the first pilot project? p. 259
3. Is the pilot program of moderate to short duration? p. 259
4. Is the pilot program meaningfully large but not overwhelming? p. 260
5. Are we able to influence the pilot program from the very beginning? p. 260
6. Do we provide pilot teams with both motivation and skills? p. 262
7. Do we let the pilot team manage the team? p. 262
8. Does management resist upward delegation by the team? p. 262
9. Are mistakes viewed as a source of learning? p. 263
10. Do senior managers devote enough time to the project? p. 264
11. Is senior management time spent in small, frequent chunks? p. 264
12. Are the new techniques communicated by direct experience? p. 265

APPENDIX A

The Role of Computerized Development Tools

Up to this point, we have focused attention on management techniques for shortening development cycles. Technology provides us with additional useful tools to accelerate development, so we address these here.

Most companies employ a combination of technology tools and management tools, but seldom do they systematically integrate these so that they work together effectively. In contrast, leading product innovators actively seek means of employing technology tools to support the management techniques they are implementing, and they consider modifying their processes to take better advantage of the available technology.

The Technology Tools Available

The variety of technologies available today almost defies description, but here we provide an introduction to the key tools.

The core technology is computer-aided design (CAD), which exists in two main varieties: mechanical CAD and electrical or electronic CAD. Even this distinction is not pure, because many products, such as motor vehicles and industrial machinery, are primarily mechanical but have a growing electrical content. Thus, to effectively design products that are both mechanical and electrical, a company must integrate these tools to create wiring harnesses and calculate circuit voltage drops in a primarily mechanical environment.

Designers today often employ many types of analysis and simulation tools to verify their designs on the computer before producing hardware or even drawings. For mechanical design, such tools can handle stress analysis, mechanisms, heat flow, fluid flow, and similar types of engineering calculations. Sometimes, engineers wait until the hardware has failed a test before conducting such analyses, but this is often unwise, because it adds time to the schedule.

Another powerful group of tools enables designers to make parts directly from an electronic CAD file, a technique called rapid prototyping. One of many rapid prototyping tools available is stereolithography, which forms plastic parts from a photo-curable resin using a computer-driven laser. When the occasion calls for high-strength parts, the designer can send the design file to a computer-driven machine tool, which can handle a broader assortment of materials but at a loss of speed. An emerging technology sometimes allows product developers to make strong metal parts quickly by using rapid prototyping to make a mold from which metal parts can be cast.

The last technology we will mention here is PDM, or product data management. PDM supports concurrent engineering by providing configuration management, by managing situations in which two designers are unknowingly changing a part simultaneously, by streamlining the processing of engineering change orders, and by supporting the flow of design documentation to the manufacturing floor.

HOW THE TECHNOLOGY AND MANAGEMENT TOOLS WORK TOGETHER

Although these technologies have great potential to accelerate certain portions of the development process, organizations often fall short of their time-to-market objectives. In these cases, the shortcoming is usually that the organization has not applied the management tools needed to reap the technologies' potential. To illustrate what can happen, we provide examples of three popular attributes of the technology and show how they can fail to yield a business benefit.

Ability to change the product's design quickly. This is a basic strength of CAD, especially of an advanced system, such as one that offers parametric design and full associativity. (Parametric design allows one to establish relationships among product attributes, so that design changes are reflected in predictable, consistent ways. Full associativity means that all software uses a common data structure, so that conflicts between views cannot arise, for example.) Unfortunately, management often abuses this flexibility by using it as an excuse to do less preplanning. Then, without a firm, clear product objective, the design team shifts daily to address the latest directive—or the team floats from design to design searching for what might please the customer. There is no substitute for a clear, stable product objective rooted in customer research.

Tightly integrated design and analysis tools to evaluate product function while designing. The problem here is that engineers are taught to refine designs until they are perfect; it is fundamental to the calculus they learn in their first year of college. Unless the project objectives and measures of success rank time to market higher than product perfection, product perfection will prevail when powerful design–analysis tools are available. As the saying goes, "At some point we must rip the design out of the engineers' hands." However, it is preferable to arrange motivational factors that encourage this to happen instead on a self-policing basis.

Ability to make prototypes quickly. Such technologies encourage product developers to say, "Let's just make a prototype and see what happens." However, without an objective, the prototype is unlikely to advance the design. For prototypes to fuel progress, a discipline of hypothesis testing should be in place first. Then each prototype is built to answer certain previously stated questions. A hypothesis-testing mentality shortens cycle time in two ways. First, it keeps prototypes simple so that they can indeed be made quickly, rather than exploiting the capability of the rapid prototyping system to provide realism and detail. Second, it instills in the decision makers—especially in management and marketing—a mentality of making specific decisions so that the team can move on. Otherwise, the decision makers may just compliment the team on its model and then ask for a functional or even more realistic model before deciding to continue the project, effectively negating the benefits of rapid prototyping.

Any of these technology tools can dramatically cut cycle time, but only if the project objectives are clear to all and the organization has communication, problem-solving, and decision-making strengths consistent with the technologies' capabilities.

CASE STUDIES

Now we show what five companies have done to accelerate their product development by integrating technology tools with effective process management. These firms all use Pro/ENGINEER,® a powerful and popular system that integrates the development process from concept through manufacturing, including mechanical CAD, functional simulation and support for analysis, PDM, and links to popular rapid prototyping systems.

American Crane, of Wilmington, North Carolina, U.S.A., manufactures lattice boom cranes. In their first use of Pro/ENGINEER, this firm developed an all-hydraulic crane in the 100-metric-ton class, a project that went from concept to prototype-on-the-test-pad in 18 months. This accomplishment was due in part to the speed at which parts were modeled. Parts that formerly took two or three days to produce on their prior 2D system were produced in two to three hours. And when a model needed changes, weeks of work became minutes, since Pro/ENGINEER's associative and parametric qualities adjusted all related dimensions automatically.

Engineers used Pro/ENGINEER's design optimization capabilities to assess the structural performance of that first crane, especially the boom, car body weldments, and upper structure weldments. The use of design optimization helped Ed Bens, a senior project engineer, and his colleagues avoid time-consuming design errors. It also allowed them to detect problems early, while they were easy and inexpensive to fix. When the decisive moment—the required SAE tests—came for the first crane, its performance was right where they wanted it to be.

"You want it to pass, but just barely," says Bens. "You want it to be safe, but not too costly. After the tests on the first crane, we had only two areas where modifications were needed. These were very minor, primarily stress concentrations in a small area that we solved by grinding off some sharp corners. Historically we had many surprises and costly rework after SAE testing," Bens adds. "But with Pro/ENGINEER, we caught problems at a very early stage."

With the design of American Crane's first hydraulic crane complete after 18 months, the second, larger model took only about six months. Using Pro/ENGINEER's associativity and parametric design abilities, engineers quickly altered original part models to change thicknesses or modify cross sections.

Note that Bens and his team used this technology tool in several ways to save time, not just by modeling parts faster. They used simulation and optimization capabilities to evaluate designs before building hardware, which saved fabrication and testing time later. When designs required changes or American Crane needed a new scaled-up design, the software assumed the bulk of the clerical burden.

Nova Link, Ltd., located in Mississauga, Ontario, Canada, is a producer of modular office furniture used in equipment-intensive environments, such as financial trading rooms. Tony Vander Park, who founded Nova Link, explains, "Our furniture is used in environments where there are lots of cables, computers, telephones, and odd bits of equipment that need to be packaged into an accessible and usable system. Hence, our product name—SYSTEMPAC. Every project is different, with different combinations of equipment and different operating concepts. We need to be able to respond quickly to unique customer requirements. Recently we developed a product and presented a prototype to a client in two days. The result? An immediate order to equip a U.S.-based investment bank's satellite office in Seoul, Korea."

There was one catch to the Korean order: Nova Link had to finish and ship the entire system in one week. "We had additional modifications to do," continues Vander Park, "including relocating a support feature and moving the cable way forward to a more accessible location. But it was all done, on time, with Pro/ENGINEER. This major investment bank has now earmarked us for three additional projects."

Just as with American Crane, this firm built a process that supports modifying designs quickly to accommodate changes in customer requirements without compromising the schedule. Nova Link also established a strength in turning design concepts into prototypes quickly, which is a crucial competitive advantage in its marketplace. In short, Nova Link decided just which short-cycle capabilities were vital to the company and then made them happen.

Stanley Tools is a venerable maker of hand tools, based in New Britain, Connecticut, U.S.A. Greg Wertheim, a product design engineer at Stanley, observed that well-selected technology tools could open up process opportunities way beyond just making drawings. "As we went along we realized that we're not here to create drawings; we're here to create products. A drawing is only a by-product of that."

In the past, according to Wertheim, designers were responsible for fully completing the design, completing detailed drawings, sending the drawings out to model makers, holding design review meetings with manufacturing engineers, refining the design, and finally releasing the drawings to moldmakers. They performed these steps sequentially, each starting when the one before it was completed. As a result, the process used to make a working prototype from dimensioned drawings could take up to 12 weeks, depending on the complexity of the project. With Pro/ENGINEER, Wertheim has used several service bureaus for stereolithography and selective laser sintering (two rapid prototyping techniques). These bureaus create actual working models that exactly duplicate the designer's intent in a week, a time savings of 90 percent.

In one project, the resulting prototype of a new tape rule met with widespread approval in customer focus groups. Once the engineers and managers at Stanley knew that users would be pleased with the new offering, they ordered production molds for the new product. By working with data directly from Pro/ENGINEER, the moldmaker could assure them that the design shown in focus groups would be the very same product offered to consumers.

"We've been able to shave weeks off our design time by using Pro/ENGINEER," says Wertheim. "The number of iterations we can go through, the number of changes we are able to make, it's just unbelievable. We have found that we are able to make some changes that would be absolutely impossible to calculate using a 2D package. If we need to make a change, we can go into the system, change the dimension, regenerate the model, and create a hard copy or a model of it. Then if marketing kills the project, we didn't spend weeks dimensioning drawings that end up going through the shredder."

Stanley Tools has made effective use of rapid prototyping and electronic transfer of files to moldmakers. More profoundly, it has used Pro/ENGINEER to support a move toward overlapped development activities and has found a means of avoiding the waste of documenting projects that are killed later. Notice in the previous paragraph, however, Wertheim's statement regarding the ease of making design changes. This ease of changing could create more chaos than benefit in a company where the designers did not receive clear guidance or have their feet firmly planted in the business world.

Apex Metal Designs supplies metal stampings and assemblies to the automotive industry from Kitchener, Ontario, Canada. In one project for Chrysler Corporation, they designed a reaction bracket for a transmission. Chrysler wanted a part that satisfied load and deflection requirements, and they wanted it within two weeks. For fuel efficiency, it also had to be as light as possible.

Typically, on such a project, Apex engineers might overdesign this part to ensure its strength under the time pressure, then perhaps run one finite element analysis (FEA) to verify that no high-stress areas would cause problems later. With the reaction bracket, however, Apex took a different approach. Within 20 minutes of modeling the bracket with Pro/ENGINEER, computing group manager, Harry Tempelman, ran his first structural optimization. After viewing the results, he decided to try some what-if studies to improve the bracket's performance while eliminating excess weight. Three days and 30 iterations later, Tempelman had a bracket that exceeded Chrysler's expectations.

An attribute of Pro/ENGINEER's functional simulation software that helped Apex was its use of p-element technology. "Unlike traditional FEA that requires some 'defeaturing' of a model for analysis, Pro/ENGINEER lets us model and analyze a part exactly as it appears in the real world—complete with the sharp corners left after machining, for example," explains Tempelman.

Apex Metal Designs has turned the tight link between Pro/ENGINEER's design and functional simulation capabilities into a competitive advantage. They no longer wait weeks for traditional FEA and prototyping to show how well a design functions. In one tenth that time, they can run dozens of simpler iterations, optimizing parts to remove excess metal without jeopardizing strength or rigidity.

This Apex example shows how close integration between design and analysis (and prototyping) provides a potent product development capability. To realize the benefits, however, the engineer also needs the experience to know which of these three tools will answer the question at hand the fastest (see pages 219–221). Also, the engineer must be in close contact with other parts of the organization—especially with the customer—to know when he or she has reached the subtle point of "exceeding customer expectations." The days of the analyst working off in a dark corner of the organization are gone.

Holset Engineering Company Ltd., of Huddersfield, U.K., is a subsidiary of Cummins Engine Corporation that manufactures turbochargers. Ian Lawson, their manager of engineering computer systems, explains the CAD challenges in their business: "Our product consists of complex aerodynamic shapes and free-form surfaces that challenge the geometric modeling capabilities of most CAD/CAM systems."

One complication is what Lawson calls "interpretability." "For our types of components, 2D drawings are inadequate and open to interpretation, since they are not a complete three-dimensional representation. When we produced 2D drawings of our 3D designs for pattern makers, their 'interpretability' made things very difficult for us."

With older 2D technology, Holset had to laboriously prepare six E-size drawings, then bring them to a foundry to shape the actual pattern. Now, with Pro/ENGINEER's manufacturing capabilities, the 3D solid model definition can be used to easily produce 3-axis toolpaths to machine patterns and dies—all in house. Lawson notes that with this advantage, "Pro/ENGINEER cut our design-to-qualified-part production time down from 35 weeks to only 12 weeks on two types of key components."

Lawson has found other opportunities for improvement too. "We frequently design two or three prototypes, each with a slight variation. We can really use the power of Pro/ENGINEER by redefining the features of one prototype to create the others. That process was much more complicated and much slower with our old CAD system."

He summarizes, "Perhaps the most significant advantage of Pro/ENGINEER is that it simplifies the entire product development process by enabling us to quickly and easily create one master geometric definition of the entire product—a definition that we can use for aerodynamic, stress, and thermodynamic analyses; that we can use to create a prototype; and that we can use to create our production tools and even the illustrations for our service and user manuals."

Holset is a technically sophisticated company that can take advantage of powerful tools, such as Pro/ENGINEER. Yet, the management challenges can be even greater than the technical ones in such companies. For example, engineers might be inclined to solve interpretability issues by just passing full geometric models electronically. Then the verbal communication channel can atrophy, and bigger-picture non-data issues, such as *why* changes are being made, may no longer be conveyed. Technology can only handle certain portions of the communication needs.

GETTING CYCLE TIME BENEFIT FROM TECHNOLOGY TOOLS

These five companies did not achieve better business results by simply buying and installing Pro/ENGINEER; their competitors can do that too. Instead, they took several other steps, both before and after the purchase, to ensure integration of the technology tools into their culture, operating style, and market environment.

Their first step was to do enough planning to know just where competitive advantage lies in their industry and how the technology might enable them to compete at a new level. This identified the capabilities they needed and pinpointed the high-leverage opportunities in their development process. It also helped them to identify the technology tools that would accommodate company growth and changes in the business environment.

Although engineers are the main users of these technology tools, the development process implications include production, marketing, purchasing, and other disciplines involved in product development. Therefore, leading companies do not regard these technologies as engineering's domain, and other functions actively participate in their planning and procurement.

Just as some technology tools are more sophisticated than others, some of the tools' suppliers are more interested than others in the broader issue of how the tools will improve your business. Partner with a technology provider that will support you in making certain that the tools improve your development process and fulfill your business objectives.

Take care to avoid the common practice of automating existing processes. This will further legitimatize them, making them that much harder to change to a faster process later. Notice that none of the five companies described above was satisfied with just making drawings faster; they concentrated on making real process changes.

Lastly, support the changes you are making with education. This should go well beyond just training engineers and designers on Pro/ENGINEER or other software packages. Training should be provided for all individuals responsible for bringing new products to market, not just the engineers who will use the software. Be sure it addresses the process, teamwork, and communication issues that will enable the software to provide a genuine competitive advantage.

APPENDIX B

Development Speed and World-Class Manufacturing

We are often asked about the relationship between the techniques described in this book and the assortment of programs that travel under the umbrella of world-class manufacturing (WCM). WCM means different things to different people, so let us focus here on the three most prevalent aspects of WCM: total quality management (TQM), Just-in-time (JIT), and employee involvement (EI). In addition, we will discuss two important techniques that have been gaining in popularity with WCM practitioners: quality function deployment (QFD) and concurrent engineering (CE).

WCM IS NOT A SUBSTITUTE FOR HAVING A STRATEGY

Before beginning, we must place a caveat on the whole concept of WCM. As WCM has become increasingly popular, it has sometimes come to be viewed as a homogeneous state of grace to be achieved only by some but desired by all. After all, why wouldn't anyone who manufactures want to be world class?

Being a world-class manufacturer is, however, a much more complex issue. First, recognize that there are many successful business strategies that do not require a company to be a world-class manufacturer. For instance, in the commodity business manufacturing cost is in theory highly critical. If a manufacturer of compost for fertilizing gardens sells his or her product at a price of three cents per pound, he or she may be protected from world-class competition by transportation costs. World-class competitors may be prevented from locating their plants in the service area of the regional-class compost company for a variety of reasons. For example, zoning restrictions may be imposed when the community no longer appreciates the delightful odor of maturing compost. In such a case it would be almost impossible for a world-class manufacturer miles away to achieve a manufacturing cost, quality, or service advantage that would compensate for the transportation cost differential for such a product.

The fact is that some manufacturing businesses are not global in scope, and being world class can be irrelevant to these businesses. In fact, being world class in the wrong business can even be damaging, because it takes resources to become a world-class manufacturer, which might be better spent on some other aspect of the business. Economists would say that there is an opportunity cost in expending resources to become world class. For example, the compost company might better spend its management time and dollars to expand demand in its regional market by educating homeowners about the relative advantages of compost, rather than trying to compete on the basis of its manufacturing strength.

Thus, not every sound business strategy requires a company to attain WCM status, even in the commodity business chosen here as an example. Blindly pursuing WCM is not a substitute for carefully analyzing a business strategy and determining what portion of the business, if any, truly needs to be world class.

The second problem in pursuing WCM single-mindedly is that the advantages of attaining it cannot be measured simply along a one-dimensional axis. It is necessary to decide which performance measures to be world class in. Reaching for the higher levels of performance will require making explicit trade-offs as to how far to go in pursuit of different aspects of WCM. For example, to achieve exceptional order-fulfillment levels may be incompatible with having abnormally low inventory levels. And having low labor costs may prove incompatible with maintaining a low capital intensity. Ultimately the trade-off frontier is reached where one cannot go further on any one dimension of performance without sacrificing on another dimension.

Many proponents of WCM point out that since most businesses are far from this trade-off frontier it is useful to think of it as not existing. They argue that it is possible to improve on all dimensions simultaneously, as if in a trade-off–free world. Practical experience suggests that this is quite idealistic. Companies that conclude that all aspects of WCM should be pursued at once often end up with extremely high fixed costs, due both to their capital investment and overhead spending. These costs then sap their financial strength before the promised benefits arrive.

We advocate early recognition that some things pay off faster than others and that concentrating on them first makes it possible to generate the resources to pursue many other opportunities. Each company must decide for itself what flavor of WCM it needs to achieve, then establish proper priorities to pursue it.

DOES WCM INTERFERE WITH RAPID PRODUCT DEVELOPMENT?

Let us assume that WCM is important to a strategy and that there is a deliberate plan for pursuing it. Should we be concerned about the impact of WCM on the ability to develop products quickly? To answer this question we need to focus individually on different aspects of WCM.

TQM AND JIT CAN SLOW DEVELOPMENT

TQM and JIT are closely related concepts. Both these techniques will inherently slow the development process if nothing else changes. However, the techniques described in this book can in most companies more than compensate for this slowing. The best news is that the slower a development process currently is, the more likely it is that it can be shortened with no impact on existing JIT and TQM efforts.

These new techniques inherently slow the development process simply because they shift the pressure for precise product and process definition upstream into engineering. This pressure increases the engineering work and slows the development cycle. Without substantial compensating changes, companies implementing JIT and TQM are likely to see massive slowdowns in their development cycles.

First, let's look at the effect of JIT, which uses small lot sizes to reduce cycle times through a factory and lower the work-in-process inventory. Such small lots can be processed successfully only when the parts can be manufactured with short set-up times and predictable yields. Unfortunately, it takes extra engineering work to achieve these goals. New tasks such as process capability studies must be performed, which take development time.

Not only does JIT require more engineering time, but rapid product development in a JIT environment places additional demands on manufacturing. Imperfect new designs have the potential to bog down a JIT manufacturing process that has little reserve inventory to absorb product glitches. The JIT approach is inherently less tolerant than a traditional manufacturing approach of the bugs inevitable in every new design.

Often JIT is explained by using the analogy of lowering the water level in a pond until some rocks appear, removing those rocks, then lowering the water level further to find more rocks. In this sense introducing a new product using traditional engineering practices is like dumping a trunkful of boulders into the pond—it can instantly overload the system.

TQM brings its own additional problems. It requires that we design in quality rather than inspect it in. It has been traditional to tolerate fairly low yields during the early stages of production, or to "engineer the product on the factory floor," as it was called, but this is no longer acceptable. Now we must design for a factory that is trying to achieve reject levels 10 to 100 times lower than those tolerated in the past. As with JIT, a poorly designed product will overload the system. At the extreme, TQM demands matching the product requirements and process capabilities so accurately as to produce no defective parts. This, too, is possible only with extra work in engineering.

Often, TQM and JIT efforts are linked to a drive for continuous improvement in productivity, which brings even more problems. These productivity increases are usually obtained by substituting capital for labor, which forces us to design for automation. Yet automated processing and handling have different demands than do traditional processes. They require a superb understanding of process capabilities and a willingness to design parts to a new, often unfamiliar set of constraints. Again, we simply increase the amount of engineering work that must be done. If the development process was optimized before adding this work, it can only get longer afterward.

Many proponents of JIT and TQM suggest that these techniques have no negative effect on the development cycle. Unfortunately, this conflicts with both our own observations and our practical understanding of the development process. We have even seen situations where new product development was effectively stopped when the design engineers were reassigned to the factory floor in teams to help start up JIT and TQM programs.

EI EFFORTS ARE BENEFICIAL

EI seeks to mobilize the normally underutilized lower-level workers in an organization. It recognizes that the majority of the capability in any organization, and particularly in one with trained knowledge workers, is located at its lower levels. This capability is typically used only partially in most companies where power has been concentrated near the top of the organization. EI empowers lower-level employees with additional training, information, and authority, thereby harnessing this latent capability.

The techniques used to accelerate development cycles are complementary to EI efforts. In trying to get more decision-making power at the development-team level we are using the same techniques EI would use.

The effects of EI efforts are particularly dramatic in industries with a high rate of technological change, such as electronics. In these industries management is particularly prone to falling behind technically, simply because there is not enough time both to manage and to stay abreast with technology. In such businesses the lower-level employees are usually the best informed technically, so it becomes critical to tap the full capability of these people. The same problems occur even in industries with mature technologies because it takes technical expertise to push them to their competitive limits.

We have never found conflicts to exist between a company's EI efforts and its efforts to shorten development cycles.

COPING WITH WCM

Manufacturing's request for better engineering is nothing new. They have always asked engineering to do a better job but now, with the emergence of world-class manufacturing, they have both a heightened need and the leverage to get what they want. As defects caused by manufacturing decrease, attention shifts to ones caused by engineering. As the professionalism and power of a manufacturing organization grow, it becomes a more demanding customer unwilling to accept from engineering what it would have in the past. As the organization raises its standards it takes engineering more time to meet the tougher new ones. There is nothing inherently wrong with these demands except that they can lengthen the development cycle. This is yet another example of the increased demand for engineering resources discussed in Chapter 11.

Companies typically cope with this change in two ways. First, they become less aggressive in their product introductions and deliberately avoid designs that overly challenge their manufacturing capabilities. New products thus become boring reworks of old ones. Second, they simply stretch out their development cycles and let their introduction dates glide gracefully into the future. When asked to provide better engineering, they simply take more time. Neither of these approaches is desirable.

In many cases, pursuing WCM is a sound goal, and fully exploiting the techniques described in this book can more than compensate for the potential slowing effect of WCM. But if you concentrate on WCM to the exclusion of the issues in this book, you may become a low-cost producer of high-quality products that are years behind the marketplace.

QFD IN THEORY SHOULD HELP

In addition to the standard techniques of WCM, two additional ones have emerged that affect the development process. These are quality function deployment (QFD) and concurrent engineering (CE). QFD is a rather elastic label used for a variety of decision-making techniques that rely heavily on matrices to present and analyze information systematically. Unfortunately, to define it more specifically would risk the ire of QFD practitioners by defining it too narrowly. In fact, QFD is now stretched by some practitioners to include almost every analysis technique that uses a matrix, whether or not these techniques were in use long before QFD appeared. The overall QFD framework will continue to grow, so let us apologize in advance for taking a narrow view of it.

The aspect of QFD that is most impressive is its systematic techniques for listening to the voice of the customer. QFD is a powerful device for ensuring that product features and technologies are ultimately traceable back to needs articulated by customers. By forcing the problem into data matrices and weighting factors according to their perceived importance, QFD forces methodical, substantive communication about product development decisions.

Critics argue that QFD achieves its benefits at the expense of a great deal of paperwork and that it is perhaps most useful for complex products and organizations. Proponents reply that you only end up with a lot of paperwork if you are focusing on the wrong issues, and they cite simple products that have been improved by using QFD techniques.

Because these techniques are just beginning to be used, we have insufficient experience to reach conclusions about their effectiveness. Let us point out, however, that QFD would appear to be at best a partial solution to the problems discussed in this book. It will not tell you to co-locate your teams, not act as a substitute for economic analysis, not force you to choose a product architecture optimized for development speed, and not guide you toward many of the other techniques we have been recommending. In fact, QFD could further stretch the fuzzy front end of the development process (see Chapter 3).

Nonetheless, QFD could prove a powerful technique for improving communications and ensuring that more systematic decisions are made. Both of these ingredients are important to rapid product development, but this does not make QFD a universal solution to all product development problems.

CONCURRENT ENGINEERING WORKS

Concurrent Engineering (CE) goes under a number of labels, in its broadest sense being called simultaneous management. Narrowly, it consists of the simultaneous accomplishment of product and process engineering. We have seen it work and you may well have noticed that it parallels our recommendation in Chapter 13 to develop product and process simultaneously.

CE has provided important time savings by improving communication, achieving better design trade-offs, reducing design rework, and allowing process development to take place in parallel with product development. All these work to reduce the length of the development cycle. There are no inherent conflicts between CE and the techniques we have been recommending throughout this book.

Nonetheless, CE is a broad tool that is not targeted exclusively at development speed. Its effects can be manifested as savings in schedule or cost or as improvements in product performance. Its very breadth can create conflicts in its application.

Despite its utility, we would consider CE a partial solution to the problem of shortening development cycles. It does increase the effectiveness of the engineering process, but has two main drawbacks. First, it is typically not guided by economic analysis and is thus vulnerable to poor business decisions. Without an economic framework one can easily conclude that no amount of CE is too much, although managers know that this cannot be true. In one case we saw teams of twenty representatives from a prime contractor using CE fly in for multiple coordination visits with a subcontractor simultaneously developing an important subsystem. This level of coordination is not inherently bad, but in this case the subsystems had life-cycle purchases of less than $300,000 and, based on what we saw accomplished in these meetings, we could only conclude that many thousands of dollars had been wasted. We believe strongly that all development management decisions must be treated as business decisions.

Our second concern with CE is that it insufficiently exploits opportunities at the fuzzy front end of the process. Getting a team off to a quick start, working quickly and effectively with marketing, controlling the scope of work through incremental innovation, coming up with a well-balanced product specification, and choosing system architectures suited for speedy development all tend to be underutilized in organizations we have seen using CE.

Despite these concerns we feel that CE can be a powerful tool and an appropriate one of the several available to help shorten development cycles.

YOU NEED A FULL TOOL KIT

This discussion of WCM techniques has revealed a lot about our approach to the problem of shortening development cycles. Others may start with their tool or framework and then identify how the tool can shorten development cycles. In contrast, we have started with the problem, observed what works in real situations, and attempted to describe this with comments on what appears to cause certain techniques to work. When an approach works, we add it to our tool kit, even if it does not fit into an existing framework.

Consider JIT, for example. Despite conflicts between JIT and rapid product development it is a rich source of analogies for speeding up product development. Many of the approaches used to shorten cycle times in the factory will also shorten them in the development laboratory.

This pragmatic approach leads to one final comment on the use of tools to shorten the development process. We feel strongly that the problem of shortening development cycles is complex and requires a full kit of tools. A good practitioner is constantly collecting new techniques, regardless of whether they fit neatly into somebody else's framework. The possibility of finding a universal tool to solve all development problems is remote. However, by careful observation and reflection you can assemble a kit of tools capable of solving most problems.

Index